D0078393

HARVESTERS AND HARVESTING, 1840-1900

DAVID HOSEASON MORGAN

HARVESTERS AND HARVESTING 1840-1900

A Study of the Rural Proletariat

CROOM HELM
London & Canberra

Croom Helm Ltd. 2-10 St John's Road, London SW11

British Library Cataloguing in Publication Data

Morgan, David Hoseason
Harvesters and harvesting, 1840-1900
1. Agricultural laborers - Social conditions
- England - History
I. Title II. Nuti, Frances
305.5'6 HD1534

ISBN 0-7099-1735-X

Printed and bound in Great Britain by
Biddles Ltd, Guildford and King's Lynn

CONTENTS

By the same author:

The Place of Harvesting in Nineteenth-century Village Life in Raphael Samuel (ed.), 'Village Life and Labour' (Routledge & Kegan Paul, London, 1975)

with Barbara Adams, Judith Okely and David Smith, 'Gypsies and Government Policy in England' (Heinemann for the Centre for Environmental Studies, 1975)

EDITOR'S NOTE

The main research for this book was undertaken between 1970 and 1973 at the University of Warwick. Although much of the book had been prepared for publication the work was not completed. In finishing this task I have made alterations to the arrangement of the material for originally the chapters were very long and included matter more usefully presented under separate headings.

Chapter 1, 'Low Farming to High' was part of the preface and the former first chapter. Chapter 2, 'The Harvest Scene' was Chapter 5. Chapter 3, 'Crops and Labour' (which has been re-titled), Chapter 4, 'Children's Labour and Education' and Chapter 5, 'The Irish Harvesters' have been formed from the original Chapter 4. Chapter 6, 'The Assessment of Harvest Earnings' was formerly Chapter 2. Chapter 7, 'Distribution and Methods of Harvest Payment' was Chapter 3 plus parts of Chapter 1. Chapter 8, 'Social Discontent and Discord' was Chapter 6 and part of the Conclusion. Chapter 9, 'Harvest Customs' was originally Chapter 7; followed by the Conclusion, now Chapter 10.

Some of the highly detailed material has been moved to the footnotes.

A very few linking passages have been added where the rearrangement of the material into book form has required it.

The author wished to thank E. P. Thompson for his encouragement and hospitality during the research work.

I am grateful to Mrs N. Moore who has typed the final version from an extensively reworked original.

Frances Nuti

LIST OF ABBREVIATIONS

A & P	Accounts and Papers
'Ag. Gaz.'	'Gardeners' Chronicle and Agricultural Gazette'
'Agric. Hist. Rev.'	'Agricultural History Review'
Bod.	Bodleian Library, Oxford
Berks RO	Berkshire County Record Office, Reading
Bucks RO	Buckinghamshire County Record Office, Aylesbury
E. Sussex RO	East Sussex County Record Office, Lewes
'Econ. Journ.'	'Economic Journal'
'Econ. Hist. Rev.'	'Economic History Review'
JOJ	Jackson's Oxford Journal
'Journ. of Agric.'	'Journal of Agriculture'
'Journ. Econ. Hist.'	'Journal of Economic History'
JRSS	'Journal of the Royal Statistical Society'
JRASE	'Journal of the Royal Agricultural Society of England'
N & Q	'Notes and Queries'
Oxford RO	Oxford County Record Office, Oxford
OJ	'Oxford Journal'
OT	'Oxford Times'
PP	Parliamentary Papers
'Quart. Journ. Agric.'	'Quarterly Journal of Agriculture'
'Quart. Journ. Econ.'	'Quarterly Journal of Economics'
Reading Farm Records	Accessions of Historical Farm Records, University of Reading
RLSC	Royal Leamington Spa Chronicle

1

LOW FARMING TO HIGH

An examination of the place of harvesting in the changing circumstances of nineteenth-century commercial agriculture in southern England between 1840 and 1900 also offers an investigation into the economic and social conditions of the rural population involved; while the extent of change in the traditional practices and customs associated with the harvest can be shown.

The three counties chosen for particular reference were Berkshire, Buckinghamshire and Oxfordshire, loosely described as mainly rural counties located south of Caird's line. These three counties are taken as a broad context for discussion and examination rather than as counties with precise individual circumstances, and no attempt, therefore, has been made to restrict either evidence, discussion or illustration to these counties alone.

The focus on the rural population has been directed to the agricultural workers, definitively and incorrectly described in the nineteenth century as the peasantry or the peasants, indiscriminately as the rural poor but, most accurately, though with unfortunate undertones, as the rural proletariat.

Harvesting in the third quarter of the nineteenth century was on an unprecedented scale; the tools available in the middle of the nineteenth century and those used still later were those of subsistence farming; the rural population primarily engaged in agriculture in the middle of the century was still growing despite emigration and migration to urban areas; at the same time the impact of agricultural depression in the last quarter of the century was most severely felt in the corn-producing counties. The importance of harvesting in the social and economic life of those areas in southern England primarily engaged in arable cultivation between 1840 and 1900 is shown by the fact that a greater number of the rural population were involved in it than ever before or since that period - so evocatively described as 'The Golden Age' - between the repeal of the Corn Laws (1846) and the agricultural depression of the 1880s.

The full implication of this situation can best be made explicit by direct reference to specific agricultural circumstances of the particular region under study and to the reasons for changes in the practice of nineteenth-century, mainly commercial farming in a wage economy from those of the previous century's mainly subsistence farming. Caird writes: 'Before the discovery of artificial manures the rich clays were almost the only wheat growing lands in the United Kingdom.'[1] If, in the period between 1800 and 1840, an increase in corn production in southern England was achieved through the reorganisation of arable farming and the adoption of new methods of cultivation, it did not occur without serious social and economic consequences for the rural populations of the purely agricultural counties.

Enclosures increased land usage, but they drastically curtailed access to land and the common rights of the rural poor; if new crop rotations increased efficiency, they led to a decline in work opportunity in the slack winter months, a situation implicit in the circumstances of larger scale arable farming. While increasing urbanisation provided a growing commercial market urban industrialisation accelerated the decline in rural home industries which had previously provided alternative family employment. In addition, wages were inevitably low in arable districts where the absence of large towns and chances of other occupations kept down the wage rate.[2] 'Low' farming,[3] however, with its reliance on the cultivation of more land through either the enclosing of waste and the breaking of marginal land, or the elimination of fallow in the rotations, neither produced cheap bread grain under the protective clauses of the Corn Laws, nor a buoyant agriculture which might have ameliorated conditions of employment.[4] So 'the detrimental changes in the condition of the labouring classes... which the lapse of years [had] wrought by gradually shutting them out from all personal and direct interest in the produce of the soil and throwing them for subsistence wholly and exclusively upon wages' continued, while 'the natural remedy', the provision of allotments as yet in its infancy provided little alleviation, 'for the difficulty of procuring land was opposing a continual obstacle'.[5]

The opinion prevalent amongst the labourers in 1849 was that 'they could not be worse off anywhere than they were... Both the horse and the slave are fed even when... idle... when a poor wretch is prevented for even half a day by heavy rain, from working, his wages are stopped... his family [consigned] to want for the day....'[6] Somerville earlier in the decade was 'shocked at the extreme depression [in Buckinghamshire] under which each family, each principle of independence, each feeling of humanity, struggled. Irregular employment, family discomfort, female prostitution, drunkenness, idle habits, gambling, absolute ignorance, and in many cases, starvation almost absolute....'[7]

To reverse the growing rural stagnation and distress, it needed the persistent advocacy of high farming and the spur of the repeal of the Corn Laws gradually to have its principles of controlling hedges,[8] clearing ditches, land drainage, and the use of organic and inorganic manures adopted. The interest in and concern for improvement in soil management at this period is reflected in the content of the papers read to the Farmers' Club. Some of the subjects covered were manures (November 1844), surface and draining (January 1845), the advantages of breaking up inferior grasslands (February 1847), the best mode of draining strong clay soils (March 1848), high farming (June 1850), the use of guano

(February 1852), the relative value of artificial manures (April 1856), and the management and application of farmyard and artifical manures (February 1857).[9] Practical experiments too on the growth of wheat were becoming the vogue,[10] an example being those carried out by Dr Voelcker on different top-dressings.[11]

Now, however, the magnitude of the subsequent expansion created a fresh problem;[12] for the hand tools and techniques of harvesting, hitherto very little further developed than those of subsistence farming, were to prove inadequate even with an increasing labour force. Harvesting remained the dominant feature of concern until the turn of the century. In the middle years of the century summer working, especially the harvest with its massive need for hand labour,[13] had become the main economic support of the rural family in wage-structured seasonal occupation while the social attitudes and customs surrounding the harvest were rooted in the past of a vanishing society. Any reorganisation was bound to have wide social and economic consequences. A study of harvesting cannot be neglected if there is to be a fuller understanding of Victorian rural life.

Not only did country sounds change when the sickle or the reaping hook or other hand tools gave way to the mechanical reaper and eventually to the reaper-binder; the enclosure movement in the earlier years of the century patterned the land with new confusion of hedge and ditch.[14] In turn, as new methods of manuring and soil treatment developed, enabling the growth of wheat outside the traditional rich clay areas, more fallow and pasture gave way to corn production, and the fields took on a new colour in high summer. Previously barn and stackyard had accommodated the yields both of a peasant subsistence and an emergent commercial farming; now the growing magnitude of the harvests, as much due to the increase in the bulk of straw and in grain yields as to increased acreage,[15] made it necessary to build serried ranks of hay and corn ricks in the open fields, altering the autumn and winter landscape - brief monuments to the labourers' toil, substantial evidence of a farmer's local standing. But these were to be the least of the changes.

NOTES

1 J. Caird, 'High Farming Vindicated and Further Illustrated 1850', 2nd edn. (1850), Oxford Union Library Pamphlet Collection, vol. 26, p. 17.

2 John Percival, 'Wheat in Great Britain' (1948), p. 60. Also figures for wheat area in England in 1870: seven eastern counties 621,621 acres, three north-eastern counties 623,528 acres, five south-eastern counties including Berkshire 425,245 acres, seven east-midland counties 406,779. But seé graph p. 12 for comparative wheat acreages 1866-80.

3 For the arguments in defence of low farming see D. C. Moore, The Corn Laws and High Farming, 'Econ. Hist. Rev.', Ser. 2,

ACREAGE UNDER WHEAT 1868-1880

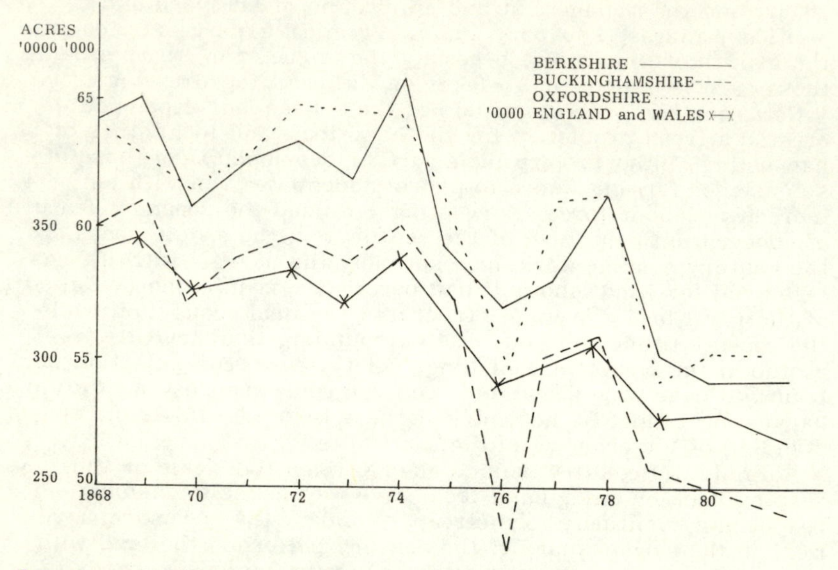

Sources: Agricultural Returns
A Century of Agricultural Statistics 1886-1966 (HMSO 1968)

vol. 18, no. 3 (Dec. 1965), p. 544: 'high farming could never be a substitute for high price... could not be so widely adopted except by subverting the traditional structure of rural society'. S. Fairlie, The Nineteenth Century Corn Law Reconsidered, 'Econ. Hist. Rev.', Ser. 2, vol. 18, no. 3 (Dec. 1965). 'A situation in which the Corn Laws protected the British farmer against continental post-war glut was giving way to one in which their retention threatened Britain with famine.'

4 Cf. 'A Century of Agricultural Statistics 1866-1966 (HMSO, 1968), p. 81. Corn returns and prices: wheat 15s 10d per cwt (1820), 15s (1830), 15s 6d (1840), but 9s 5d (1850), 12s 5d (1860), 10s 11d (1870), 10s 4d (1880), 7s 5d (1890), 6s 3d (1900).

5 PP 1843 (402) VII, Report from Select Committee on Labouring Poor (Allotment of Land), pp. 203-5.

6 Supplement to the 'Morning Chronicle', 24 Dec. 1849, p. 6, col. 4. The rural districts: Bucks, Berks, Wilts and Oxon, Letter 3.

7 A. Somerville, 'The Whistler at the Plough' (Manchester, 1852), p. 17. 'We as have apples and income to afford flour may have pies and puddings both, but every family - nor the half nor the quarter - have not fruit of their own, and if they had, where be the flour to come from and the sugar.'

8 J. B. Spearing, On the Agriculture of Berkshire, Prize Essay, JRASE, vol. 11, p. 3 (1860): 'timber and corn cannot be grown together to advantage... hedgerows are a great bar to... improvement...'.
9 Kevin Fitzgerald, 'Ahead of Their Time, A Short History of the Farmers' Club' (1968), p. 202.
10 J. B. Lowes and Dr J. H. Gilbert, Report of Experiments on the Growth of Wheat, JRASE, vol. 23 (1862), pp. 31-4.
11 Dr A. Voelcker, Experiments With Different Top-dressing Upon Wheat, JRASE, 2nd edn., vol. 14 (1878), pp. 16-31.
12 John Dudgeon, On British Agricultural Statistics and Resources in Reference to the Corn Question, 'Quart. Journ. Agric.', vol. 12 (1842), p. 493: 7,000,000 acres in grain crops, 4,500,000 acres fallow. Compare 7,020,000 acres in wheat, barley and oats, official returns (1866). Dudgeon's figures, however, include rye and beans, and returns in 1866 were incomplete: 'the reluctance of farmers... was most marked in the eastern and southern counties', 'A Century of Agricultural Statistics', p. 4.
13 By the late 1850s the cost of machine versus hand cutting had become an economic factor in production. Peter Love, On Harvesting Corn, Prize Essay, JRASE, vol. 13 (1862), pp. 217-26. Cost per 100 sheaves: 10d high reaping, 1s low reaping, 1s 2d bagging, 1s 3d mowing, 1s 1d machine reaping, 1s 2d machine mowing.
14 Mabel K. Ashby, 'Joseph Ashby of Tysoe 1889-1919' (Cambridge, 1961), p. 7: 'the vast heaths and fields had been broken up into... tiny modern fields'.
15 M. J. R. Healy and E. L. Jones, Wheat Yields in England 1815-59, JRSS, vol. 130, pt. 4, Ser. A (1962), p. 574. For increase in grain yields (p. 578), general mean yield, bushels per acre: 1815, 370 bushels; 1859, 55.1 bushels.

2

THE HARVEST SCENE

> At harvest time there were crowds of reapers in the field, crowds also of women and children tying up sheafs of corn, sometimes quite often doing the heavy work of reaping sometimes the lighter but other equally necessary jobs. Everybody was involved and if something went wrong a broken trace, a cracked fellee on a wagon wheel or a horse casting a shoe, someone in the village was there and ready to put it right; and the ritual of harvest made it a truly communal occasion.
> G. E. Evans, 'Ask the Fellows who Cut the Hay' (1956), p. 137.

Descriptive writings both past and present have provided copious information on how the harvest was gathered prior to and during the nineteenth century, and have presented innumerable pictures of the activity in the fields at harvest time. But to obtain an accurate idea of the harvest scene or even to question the accuracy of these descriptions at any period it is essential to take particular account of the historical context and the circumstances of agricultural practices.

What distinguished the harvest in the middle years of the nineteenth century was not only the fact that the entire rural population and even townspeople were involved but the magnitude of that involvement. Because of the increase in corn acreage greater than ever before or since, the harvest field, with a rural population still growing, was unprecedentedly crowded. These harvests of the 'Golden Years' were the culminating point of an agricultural revolution, basically one in land utilisation and production, not in the gathering processes. It was not only the increased yield of grain but the increase in straw production that presented new problems. The period of harvesting remained the same; the time scale of growth, fruition and decay remained the same. The same vagaries of nature and weather were still outside man's control; but the cutting tools available were still those of subsistence farming. The harvest fields in the past had also been crowded but not on the same scale, and whenever the harvest had been previously in jeopardy legal compulsion could bring out the labourers from town or village whatever their trade or calling.[1] In the nineteenth century it was economic compulsion that brought the labouring population into the harvest fields but even with all the villagers together with the help of immigrant bands of harvesters farmers were sometimes to complain that their harvest need for labour could not be satisfied.

The basic problem in harvesting is simply stated - how to gain maximum yield and quality in the minimum time. If corn crops are gathered before they have ripened then quality suffers, problems arise in storage and yields suffer. If, however, ripened crops are not harvested quickly, then quality and yield again suffer. If unfavourable weather occurs there can be crop damage and the time available for harvesting telescopes. If grass for hay is cut too early yields are less; if too late it starts to seed; if it is cut and the weather turns wet all kinds of repeated operations are needed to make the hay so that it can be carted in a satisfactory condition. If still damp when stacked, heating occurs, the quality is spoiled, it can go musty and become unpalatable for stock; the rick can even catch fire. The basic problem for the Victorian farmer can also be simply stated - how to restructure harvest operations successfully in a new set of circumstances. An examination of the tools and tasks of harvesting allows an assessment of the extent of adaptation or development in techniques or changes of practice.

While the nineteenth century was an era of agricultural expansion, it was also still an age of traditional husbandry. The methods at harvest time, the tools employed, what little machinery there was were very little different in 1850 than in 1750. With the single exception of the threshing machine, only by the end of the century did machinery begin to have any real impact on either production methods or the working lives of country people. In harvest time the work was physically hard: continuous effort and 'many hands make light work' was not an exhortation but a necessity, for it was never a case of making work light but of getting the work done at all. Four basic hand tools were the cutting tools in use: the sickle, the reaping hook, the fagging hook and the scythe. The sickle was a light crescent-shaped serrated-edged hook beloved by Irish harvesters;[2] the reaping hook, its smooth-edged counterpart, the preferred tool of the English.

> The reaper with a sickle (or a reaping hook) passed the curved blade round a portion of standing corn and drew it towards himself; the pull made the ears cluster in a bunch which he gripped, and severed the straws immediately afterwards. Tucking the corn under his left arm he repeated the action until he had enough to make a sheaf. Reaping by sickle... served the needs when storage in barns and threshing by flail was general.[3]

When corn was cut 'high' it left a long stubble to be mown and raked up later for litter when the demand on labour was less pressing, for housing the corn safely was the main concern, and small sheaves lightened the task of carting and occupied less barn space. Sometimes it was left until touched by early frost, then merely beaten down by heavy sticks or poles to save labour, with sheep put in to break the straw ready for raking.[4] When straw, whether wheat or rye, was needed for thatching, 'low' reaping was adopted.[5] Debate on the advantages of 'high' and 'low' reaping occurred as late as 1860, suggesting that in no sense was reaping yet considered outdated. Definitive reaping demanded the use of either the sickle or the reaping hook, though both could be used as a straightforward cutting tool, but the term reaping did not then apply. The fagging hook however was precisely this. Variously described as a vagging, bagging, or 'broad' blade hook, it was a much heavier, more rounded tool with many slight variations in actual design and an improved cutting edge after the 1850s (Bessemer process).[6] In fagging, the corn was untouched by hand, tensioned by a hooked stick and cut with a strong swinging slash, severing the standing corn close to the ground. This was a faster, less careful method than reaping and was variously termed 'fagging', 'vagging', 'bagging' and 'swopping'. 'Baggers', 'vaggers', 'faggers' and 'swoppers' never reaped.[7] It is necessary to make this distinction for in many contemporary nineteenth-century reports and subsequent accounts there exist discrepancies between the name of the tool used and the actual mode of cutting. Julyan presents a puzzling example in his description of the commencement of harvesting in East Sussex:

> A gang of men called swoppers went from farm to farm... to cut the corn. Men started in a line... bending down, each man would catch hold of an armful of corn laying it across their thighs, they would cut off with their swop hooks...

Either he witnessed an unusual and dangerous form of 'fagging' or merely provided a faulty description of 'reaping'.[8] More usually, however, it is the loose use of 'reaper' and 'reaping' that causes most confusion. Jefferies makes this very point:

> The labouring men used to tell me how they went <u>reaping</u>, for although you may see what is called reaping still going on at harvest time it is not reaping. True reaping is done with a hook alone and the hand; all the present reaping (sic) is vagging with a hook in one hand and a bent stick in the other, and instead of drawing the hook towards him and cutting, the reaper chops at the straw as he might an enemy.

In the same passage he makes this very error and is defining the fagging hook as an enlarged form of reaping hook - in fact an enlarged sickle:

> The reaping hook endures and is used on all small farms and to some extent on large ones to supplement the work of the machine... In itself the reaping hook is an enlarged

> sickle... the oldest of old implements - very likely it was made of a chip of flint at first and then of bronze and then of steel (sic) and now at Sheffield or Birmingham in its enlarged form of the vagging hook.[9]

Here there is a confusion between the continued use of the reaping hook on small farms and the hook used as an ancillary tool on larger farms; the reaping hook could be used in 'fagging' (though not as efficiently) but not the 'fagging' hook for reaping; the strict distinction between sickle and reaping hook was in the presence or absence of a toothed edge, while the fagging hook, ostensibly at first glance an enlarged reaping hook, by becoming a vagging hook changed its function. In losing its lightness and to some extent its shape by becoming more rounded true reaping with the hook became impossible. The fagging hook in fact was developed precisely because weight rather than lightness was needed in the cutting process, the reverse being the case for reaping,[10] while the increased curve of the blade helped gather loosely cut straw. The necessary adjunct to the fagging hook, the hooked stick, was often cut straight from the hedgerow, but sometimes it was specially manufactured of iron - the pick thank. Sometimes it was called a reap stick and sometimes merely referred to as a 'hook', suggesting two cutting tools in simultaneous use.[11] Both were contradictions; both could lead to false conclusions. When cutting peas and beans 'hacking' or to 'hack' was the term used and though special small pea or bean hooks were manufactured,[12] quite often the fagging hook was used and essentially the method of cutting was fagging. Peas and beans were also mown with the scythe but this long, broad-bladed, two-handled tool with its wide-sweeping cut was essentially the only grass-cutting tool in hay time.

In a variously developed form and fitted with a cradle to facilitate sheaving the 'corn' scythe was to prove its worth in speeding up the cutting process at harvest time and held its own against the mechanical reaper until almost the close of the century. Its use in the wheat fields at the beginning of the century was exceptional though it was often used to cut barley and oats. Least probable witnesses to its function and adoption as an all-purpose corn-cutting tool were the Irish who were to scythe their own national harvest in the late 1860s though seemingly wedded to the sickle in the 1840s.[13]

Despite the fact that mechanical reapers of many kinds were invented in the earlier decades of the century it was not until the late 1850s and early 1860s that they came to be considered as alternatives to hand-tool cutting.[14] The period in England between the Great Exhibition, which focused serious interest on mechanised reapers, and the turn of the century, when reaping by hand became an anachronism in commercial agriculture,[15] must however be seen as much a period of a switch in hand-tool usage, as a period in which mechanisation partially and in no sense completely supplanted hand cutting. The broad generalised picture might well be that by 1870 the mechanical reaper had become an important adjunct to hand cutting; but in 1900 hand cutting still remained a very necessary adjunct to mechanical cutting, with the reaper-binder pressing for acceptance in the circumstances of an overall general decline in rural population. It might also be suggested that the reaper in its earlier stage of adoption, not so much aided farmers generally in the field as influenced farmers to reconsider the whole of traditional farming practice and appraise afresh their existing range of hand tools. Drawbacks to mechanical reapers were many - they were liable to break down, they consistently could not deal with heavily laid crops, they often could only be worked one way, down or up and rarely across heavily ridged and furrowed land, they represented capital investment, and if by the late 1870s improvements in design had converted farmers to favour their adoption, their pockets kept them in two minds.[16]

If the mid-period year 1867 is taken as a guide, it is clear that harvesting practices were in a state of transition. In the early years of the century the general practice was to reap wheat, sometimes the oats, rarely the barley while concurrently mow the barley, sometimes the oats and rarely the wheat. By the middle years of the century a remarkable change in general practice and hand-tool usage had taken place. It might be suggested that it was not the labour force that had become inadequate but a labour force that was working in an out-of-date framework of harvesting techniques, inadequate to the new situation of continually rising output.[17] This is supported by the fact that change was general and not only in the counties of declining rural population.

In Oxfordshire, for example, barley and oats were no longer mown and always carted loose, but 'latterley' all tied in sheaves; in some parts of the county 'most of the crops' were 'fagged' at various prices from 10s to 14s an acre. 'Fagging has become very general in this part and many good hands are to be found, the last wet harvest has made the farmers anxious to sheaf all their corn... when properly done it will stand much wet without damage'. 'Fagging most general. A considerable quantity of the Barley is fagged and nearly all the heavier crops of Oats.' 'Mostly by fagging. I tie up all I can. The few fields that are not laid will be cut with the Samuelson's sheaving reaper, and tied by the acre [but] the corn grows very frothy rather weak in the straw so that it is difficult to cut by machinery'. 'Reaping machines are used but little....'[18]

The general impression from such reports is concern for the quality of the harvest, the general dominance of the fagging hook as speedier than the sickle, but no general concern to speed up work by exclusive use of the scythe and certainly no general interest in the mechanical reaper. The report from Buckinghamshire shows a considerable fluidity with a full interchange of the three main cutting tools: 'Reaping and mowing per acre... Wheat is reaped, fagged and mown according to circumstances; tied and set in shocks. Barley is mown and carried loose. Oats generally fagged and tied, or mown and tied. Beans are mown and crooked or fagged and tied according to length of straw and bulk of crop; a great many Bean crops this year cannot be tied.' 'Some reaped, some mown, some cut by machine and some fagged.' 'Barley fagged and sheaved, 10s per acre; but layered crops vary from 14s to 25s per acre.' 'Fagging has now become prevalent... not many reaping machines... none has been seen yet that will cut a layered crop perfectly.'[19] The reports from Berkshire, however, show a more settled situation. There is no mention of the reap hook in use; barley which 'used to be mown' is still mown but 'fagging this is, however, becoming more general every year'. The reaping machine is very little used; the fagging hook in fact has become the dominant cutting tool for all crops.[20]

In Oxfordshire's neighbouring county, Warwickshire, despite the fact that many reaping machines were used, wheat was reported as 'chiefly cut with sickle' while 'reaping and fagging and mowing' were widely practised.[21] In another county adjacent to Oxfordshire wheat was still reaped or cut with the 'heavy hook' close to the ground - the general situation in Gloucestershire was in fact described as 'Wheat about one fourth cut by machine, two fourths [cut] by scythe, one fourth reaped. Barley and Oats mown with scythe.'[22] In Bedfordshire the fluid situation existing in neighbouring Buckinghamshire was present, wheat being reaped or fagged or mown, barley still mown and carried loose with oats generally fagged and tied, but there was a general absence of reaping machines.[23] So also in Wiltshire, where wheat and oats were mainly still reaped, and barley still mown, but all at very low prices, 8s to 9s for reaping, 2s per acre for mowing.[24] So also in Middlesex: here all the corn was 'fagged and shocked' in the usual way, but prices were high - 15s being the average price for wheat and oats, and 12s per acre for barley if also fagged.[25] In Northamptonshire wheat reaping was reported as 'fast being superseded by the machine and scythe', yet others reported the old practice of reaping wheat and mowing barley loose.[26] In Cambridgeshire, further north, reaping was still practised alongside the scythe and reaping machine, with the significant absence of the fagging hook.[27] Far south on the Berkshire border, the Hampshire practice was to mow the bulk of the corn; surprise was shown that the reaping machine was not in more general use. 'To those who understand the proper working of them they are invaluable.' A significant indication of a change in harvest practice is the mention that 'much of the corn here is stacked in the field', and the distinction between tool usage - 'On the clay lands the bagging hook is used, on the hills the scythe and reaping and mowing machines' - recognises the deficiencies of the scythe and reaping and mowing machines on heavy crops subject to being laid.[28]

These general reports indicate at the very least a considerable rethinking in harvest techniques. Old methods have changed but not disappeared, yet practices may well have only pertained to the larger farms. Though the use of the sickle is hardly mentioned and the reaping hook was being generally superseded,[29] they were both still the ideal tool for cutting wheat if thatching straw was needed - and thatching straw was indeed greatly needed.[30] Thus the continuous existence of the slow sickle alongside the fast-cutting scythe is understandable and if labour shortage was a consideration then women could often reap as well as men.[31] Though Irishmen were cutting wheat with the scythe in North-East Oxfordshire in the 1880s, women were said to be reaping in the same harvest,[32] and in the districts around Burford, as late as the harvest of 1889, it was reported that 'everywhere around... the sickle had been brought into requisition and one field after another of ripened corn had been laid low'.[33] It was however a wet season. Despite the reported dominance of the fagging hook in Oxfordshire, reaping hooks were on sale in Thame in 1870, and still obtainable in the Windsor area in 1886.[34]

All the indications suggest that if machine reaping was gaining ground south of Caird's line in the late 1860s its general adoption was to be slow rather than fast until the turn of the century. If the farmers who viewed the demonstration of reaping machines at Leamington in August 1876 had any lingering prejudices against the scythe as 'a rough and ready way of getting through the work [which] involved a considerable sacrifice of grain',[35] the shock of seeing a Rochester 8 hp road locomotive with a Crosskill reaper and a cut twelve-feet wide mounted in front of the engine, charge into a piece of bright-strawed wheat must have dispelled them, and consolidated the use of the scythe certainly as the best hand tool for speedy cutting; rather than induce the purchase of such a monster or even a slightly less frightening reaper. It might not seem to have been the case for at a trial of reaping machines at Leamington upwards of thirty models were competing. At Banbury, indeed, in 1872, the Britannia Works were said to be at their peak turning out the 'remarkable number of 3,000 reaping machines'.[36] But the situation is put in perspective by G. E. Fussell who points out that 'there were 5,672,000 holdings in 1882 and 10 years later rather less than 1% had a reaper'.[37] Indeed the depression in agriculture had so greatly affected the Great Hasely Iron Works by 1893 that John Venables Gibbon, its owner, an inventor of several agricultural machines, committed suicide.[38]

The strictures on time were slightly eased in the middle years of greatest production by cutting before the near-ripe stage.[39] The reaper later slightly changed the work pattern by speeding up the cutting operation: by partially eliminating hand cutting it compensated for the gradual decline in the migrant labour force but it left unaffected the continued need for local labour. So long as labour remained cheap and available there were good reasons for postponing the adoption of the reaper. When, in the 1890s, labour was becoming 'absolutely' scarce the logical progression was the adoption of the binder, though a farmer in 1898 did not think this yet to be the case in mid-Oxfordshire:

> I have been many times urged to buy some binders... [but] there does not seem much saved by a binder except in districts where labour is scarce. No doubt the binder is the only harvesting machine of the future but it should be made to twist its own bands and so save the cost of a string.[40]

Only at this stage in the last decade of the nineteenth century in the development of harvesting practices did anything approaching true mechanisation emerge.[41]

Now there were to be direct economic and social consequences for rural workers. The switch in hand tools had meant the learning of new skills but with no economic consequences. The reaping had reduced some of the arduous work of hand cutting but had not lessened their harvest earnings. The general adoption in the new century of the binder, however, not only reduced family earnings but altered the pattern of employment, striking at the independence of village life, for more and more of the work at harvesting was to be confined to the hired men. The choice of uncertain freedom was now threatened for the opportunity of harvest work began to rest on accepting a hired dependence.

A complicated division of labour existed in successful harvesting. Cutting commenced a process which created a chain of procedures which ended only when the last bolt of straw had been carried to the thatcher and the last rick made safe. Cutting could be occurring in more than one field and the crops being dealt with simultaneously at varying stages in the harvesting process. Making bands, tying and shocking the sheaves, pitching them to load wagons and pitching them again to the rick builders were the essential links in the procedure. Wheat might be ready to cart in a matter of days; oats it was considered should stand in 'stook' three Sundays.[42] Oats, if 'mown', and barley would have to be turned like hay in the swathe for oats and barley, unlike wheat, retained sap in the straw and carrying had to wait on wilting, especially in the case of undersown barley.[43] No single group of workers could perform all these tasks, for each task for each crop had its place in a time continuum and the optimum time to perform them inevitably clashed. It was only exceptionally that individuals cut and made bonds and shocked the sheaves on their own; it was less exceptional that 'bands' of harvesters contracted to take the whole of the harvest (especially in East Anglia), but generally it was a mix of itinerant harvesters or piece workers from the open villages undertaking the cutting and being responsible to the 'shocking' stage, or merely taking the cutting and moving elsewhere to another farm for further cutting, with the 'own' casual labour being responsible generally under the carter for carting and stacking the corn. But individually or in groups there was a continuous succession of work for all available hands - women and children helping to make bands, and binding and stooking and generally helping the men; individual women reaping so long as the practice survived; men young and old cutting and carting, with women behind them raking the ground bereft of sheaves; the older rather than young men on the waggons loading and helping in building the ricks, in deference to their experience and in consideration for either their failing strength or rheumatic joints. The younger men pitched to waggon and from waggon to stack, for this was regarded as the most strenuous work, and generally the young lads and often small boys were in charge of leading the waggons.

Not only the extent of the crops but their variety made inevitable this vast concourse of workers, for the work had to be contained within a constricting time scale and with a continuing immediacy. The cut corn could not be left unsheaved and unshocked overnight; the shocked corn could not be left uncarted when ripe for risk of shedding. 'A field of wheat left uncut a day too long may have two bushels an acre blown out of it by a high wind... barley left uncarted... changed from good malting sample to cheap feeding stuff.'[44]

Cutting before the near-ripe stage slightly eased the time factor; the reaping machines speeded up the cutting process. But it seems a mistake to think that the need for labour became less. The contemporary opinion was 'that machinery in the field does not reduce the number of men employed. But they are employed in a different way. The work all comes now in rushes - acres are levelled in a day [and] the cut corn demands immediate attention.'[45] It was not uncommon when the machines first came into use to let the men use them without any charge, the farmer getting his advantage in the increased speed in securing his crops.[46] It also seems a mistake to suggest that a direct saving in labour costs occurred. 'Previously', states E. L. Jones, 'a farmer had to pay his harvest gang to sit idle in the barns while it was raining unless he was to risk turning them away and having no reapers ready for a bright spell.' Harvest gangs, however, were under contract for a set acreage of work performed or on direct piecework rates, so it was the harvesters who lost working time not the farmers who paid for idle time.[47]

Though at hay time there was only a single crop, fields cut at different times needed continuous attention, spreading and tedding the swathe, raking the swathe into rows (windrowing), and once the hay was considered made, there was still carting and rick building. The time factor was never so pressing as in the corn harvest for rain could not only prove more damaging than to shocked corn but several days work might be wasted, and often only after repeated forking and raking could the hay be made fit to carry. Whatever the terms used for the tools and the techniques at hay time and harvest, there was a general uniformity in the sequence of tasks and in the way the work was performed. Hay forks might be 'prongs', sheaves tied with bands or bonds, shocked or stooked, 'isled',[48] 'piled' or 'clumped' (Somerset), 'sticked' (Dorset) or 'thawed' (Suffolk), put into small 'hatlocks' (Cheshire) of eight or ten sheaves or into 'mows' of twelve (Shropshire).[49]

The ricks might be round or oblong or square, the waggons

lumbersome or lightened by chamfering into the elegant lines of the 'Woodstock wain' and the broad-bladed hook 'fagged' or 'vagged' or 'bagged' or 'swopped' the standing corn; yet the essential techniques of harvesting extended back in time and crossed the country. All the articles needed for sharpening the cutting tools - the stricle made of froughy, unseasoned oak, the hammer to pit it, the sand and the grease-horns used in Yorkshire in the seventeenth century and before[50] - were there in the several counties of Britain in the early nineteenth century to be compounded into the later single sharpening stone (the rubber). The dialect used for harmonious loading varied from county to county but the method was the same.[51] Over the Oxfordshire border into Warwickshire at Ilmington in the 1890s at the hay cart:

> [W]hen the loader had filled the bottom of the wagon he would say 'Corners in front'. They would then put their forkfuls up, both together, and he would hold them by putting his fork across both. 'Between corners' meant one of them put his forkful to bind the corners. Next 'Behind the corners' and so on towards the back. Then 'Corners behind' and 'Between corners'. Then 'up' in the middle to bind the load. When they wanted to move forward, they would shout 'Hold you' to warn the man on the top... They took pride in their loading and would say of a good load 'That will ride safe'.[52]

Back in time two and a half centuries, and four counties away, there was little difference:

> In loadings of a waine they first fill the body, and then doo they beginne with the far fore-nooke, and after that with the near fore-nooke, then with the far hinder nooke and last of all with the neare hinder nooke; layinge on usually three goode covers and seldom any more, for making her too high; yett some will laye on fower.[53]

'For three weeks or more during harvest the hamlet was astir before dawn... For a few days or a week or a fortnight the fields stood "ripe unto harvest". It was the one perfect period in the hamlet year.'[54] Across the north-east border into Warwickshire the feeling was the same:

> One of the greatest sights was a field of wheat being cut. The farmer would take on as many families as he could, and it was beautiful to see them stretched across a twenty acre field, each taking their allotted acre or two as their part in getting the harvest gathered. The father slashed away with his cutting hook, in his other hand using a wooden hook to hold the wheat in position. One of his school-age children would be making bands from a handful of wheat straw, and

> laying them to receive the sheaf. The mother would come later from home with a basket of food.[55]

Across into Buckinghamshire about Thame:

> Whole families planned for work in the harvest field: the father with the eldest children, to go daily in advance; the mother, with those younger, to follow later, with provender for the day. The work was done by the piece; it was a matter of slaving from morn to late evening, an incessant 'slash, slash', with the gleaming fagging hook at the ripe corn; and then the gathering of the severed halm together and the binding of it with straw band laid ready on the ground by one of the children. The work was called 'fagging', a truly descriptive title, in contrast to the earlier method of cutting with a sickle, which was called 'reaping'.[56]

'[A] cornland village is always the most populous, and every rood of land thereabouts, in a sense, maintains its man', wrote R. Jefferies in the 1870s, and continued:

> The whole village lived in the field... The reaping and the binding up and stacking of the sheaves, and the carting and the building of the ricks, and the gleaning, there was something to do for everyone, from the 'olde, olde, very olde man', the Thomas Parr of the hamlet, down to the very youngest child whose little eye could see, and whose little hand could hold a stalk of wheat.[57]

J. G. Cornish recalls that in Berkshire in 1883 when harvest time arrived:

> most of the villagers migrated to the corn fields; nearly all the wheat and most of the barley was cut by hand. The fields were usually about one furlong wide and perhaps there might have been eight acres of wheat ready for cutting and let us say eight men engaged on the task. They would draw lots for their strips in the field and then eight families would begin to drive their pathway into the standing corn. I said eight <u>families</u> though perhaps some of the men were 'widow men' i.e. bachelors would work alone and so be outdistanced by their more fortunate neighbours.
>
> Father with his broad blade fagging hook in his right hand and crooked stick in his left, threshed through the yellow stalks and left them gathered by his foot. Mother followed, swept a sheaf together, placed it on the bond, drew this tightly and fastened it with a twist. The children pulled the bonds, the younger perhaps only able to select six or eight stalks needed to make one, while the elder made it ready for the mother to use.[58]

'The harvest was the goal to which the main efforts of the past year had been directed; its ingathering was all-important to the village,' states Walter Rose, thinking back to the 1870s, and continues:

> The laying down of arable land to grass began when I was a boy; yet corn was still so widely grown that at harvest it was impossible to escape the harvest redolence that pervaded all the district. The horse-drawn reaping machine had appeared, with its four spreading sails that dropped obliquely in succession to sweep a sheaf of cut corn aside. Its use had eased the growing problem of hand labour... but as long as labour was available at an economic price, the farmers did not favour machines. Cutting by hand was clean and effective...[59]

Thomas Hatcliffe of Worksop, writing in 1905, described the general situation as it was in midland villages in his childhood 30 years earlier:

> every man, woman and child went forth into the fields to help... and win the extra wage for harvest... When the first corn field was ready... the sicklemen or scythemen with the gatherers and binders were at the field. The gatherers of the sheaves and the binders were generally the wives and children of the men, and the whole work of the harvest was of the nature of a family outing... though a hard working one... the reapers or mowers fall in one by one behind the leader, the women and children as gatherers and binders following in their wake. The first stop was when the leader wanted to sharpen. He said 'Now', and all stopped at the end of his sickle or scythe swing. Then came the music of half a dozen tools sharpening as the stone rasped the steel blades... The sharpening was often as not the time of 'lowance' as well, when from the wooden kegs or stone bottles came the welcome 'guggle guggle' of the home-brewed as it fell into the horn ale-tots... Those ale-tots of horn were held to be the best for harvest drinking... being cooler and sweeter than in any other form and far before that of 'sucking the monkey' as liquer drunk from a bottle was called. The tots were emptied at a drain.[60]

The editor of the 'Cornhill Magazine' thought 'Harvest' a pleasant subject for Christmas reading. The anonymous author continues his description after the reapers and mowers grow scarce:

> As the days go by, those groups grow scarcer, and are replaced by the heavy creaking waggons, piled high into the air with sheaves, accompanied by two or three boys, hot and shouting, a more staid-looking rustic by the side of the horses, and another, probably lounging at his ease

> upon the load. The very horses, at such times, as they strain up the last bit of hill before they reach the village, seem under some exceptional excitement, as if they too were conscious of the good time... then what a scene of vigorous active work, of rustic 'chaff' and geniality, is the stack-yard! The waggon is soon drawn up alongside the fast-rising rick, the horses are taken out, and sent back with an empty wain to the field; and then begins the process of stacking, and what is technically known as 'pitching'. The men who stand upon the stack, adjust the shocks as they receive them, and two men stand below in the waggon to 'pitch' them up to their companions. This work of 'pitching' is supposed to be the hardest of all, and is generally done under the eye of the master, who frequently plies a fork himself, just to keep his men up to the mark.[61]

A shepherd's wife, who herself used a sickle, describes working in the Sussex harvest fields in 1875:

> After an early breakfast, I used to start with my children for one of the Hall fields, carrying our dinner with us... Even the toddlers could help by twisting the straw into bands and also by helping me tie up my sheaves. But cutting and binding our sheaves was not the end of our day's work. We still had to make the shocks. Many is the time that my husband came round to us when his own day's work was done, and we worked together setting up the shocks by moonlight.[62]

Because the labour was cheap, much corn was also still cut with the sickle about Tysoe at this period. Merriment underscores Mable K. Ashby's harvest scene, but her description reveals the unusual feature of men helping the women, only ever associated with reaping:

> A very neat small instrument it was; a good worker, dropped hardly a straw where the corn stood up well. A dozen or maybe 20 reapers largely women would work in the field with men following up to tie the sheaves and another group to set them in shucks... there was talk and banter and flirting and yarn spinning during the meals under the hedge.[63]

Jefferies, writing of the same period, emphasises the work not the relaxation of harvesting:

> One evening there was a small square piece cut at one side [in a field], a little notch, and two shocks stood there in the twilight. Next day the village sent forth its army with their crooked weapons to cut and slay... More men and more men were put on day by day, and women to bind the sheaves ... as the wheat fell, the shocks rose behind them, low tents

> of corn. Your skin or mine could not have stood the scratching of the straw, which is stiff and sharp, and the burning of the sun, which blisters like red-hot iron. No one could stand the harvest field as a reaper... Their necks grew black, much like black oak in old houses. Their open chests were always bare, and flat, and stark... The breast bone was burned black, and their arms, tough as ash, seemed cased in leather. They grew visibly thinner in the harvest field, and shrunk together - all flesh disappearing, and nothing but sinew and muscle remaining. Never was such work... So they worked and slaved and tore at the wheat... the heat, the aches, the illness, the sunstroke, always impending in the air - the stomach hungry again before the meal was over, it was nothing. No song, no laugh, no stay - on from morn till night...[64]

If in the 1870s the sickle was still in use in South Warwickshire where 'the old men made a very poor cut with the fagging hook: about half an acre a day', it was about this time when it was being superseded in the Cotswolds.[65] The change to the exclusive cutting of grain with the scythe led to difficulties. A young Essex lad noticed how:

> strong men work and rend the corn up by the roots, whereas the art is in knowing how to whet the scythe... The scythe should have a keen edge, fairly smooth, not too ragged; then in cutting ripe grains one must keep the left arm fairly high up. So the operator can cut twice the distance that one does who is careless in his work. It used to be a pretty sight to see 8, 10, 14 or 16 men in different fields all... cutting the grain in beautiful swathes, right across a 20-acre field.[66]

Isaac Mead, the Essex lad, was 15 years old when he took a scythe for the first time, to work in with a company of 13 others in the harvest of 1874. Arthur Randell recalls how as a much younger child he worked in the Fen harvest fields at the turn of the century:

> My parents had harvested on the same farm for many years and as soon as we children were able to do so we had to take our share in... making the bands to tie the corn and then, as we got older, to tie and shock alongside our elders.... I have often seen the harvest men during a spell of very dry weather, take armfuls of straw and lay them in the bottom of the Nook Drain to soak, otherwise the bands being too dry would have broken as soon as they make them.[67]

Another fenman was only eight years old when he first worked a full day in the harvest fields, the harvest of 1899:

> A few minutes past 6 o.c. I was taken to a field of wheat where a six sailing reaping machine, with four horses attached, was waiting... I was to ride the off-side front horse, keeping it close to the standing corn, then pull out on reaching the corner and swing back. The horses were changed at 10, 2 and 6 o'clock; there was no dinner break, one of the men taking my place while I was eating... The two men were paid one shilling an acre, so by cutting 10 acres a day, they earned five shillings each. Just as dusk was falling I ended the longest day so far in my young life. I spent three weeks riding the reaping horse, then I was switched over to leading the horse and shouting 'Hold tight' when corn carting started... Besides the two men [loading and pitching] there was also a woman who dragged a large wood rake to gather up loose falling strands.[68]

The large wooden rake would have been the heavy rake described by Best - the sweathrake:

> [A] sweathrake is soe called for that it raketh a whole mowers sweath att once; for as an ordinary mower taketh a broade lande att fower sweathes, soe doth hee that traileth the sweathrake take a while lande at twice goinge up and twice downe. A sweathrake hath usually 33 teeth, sometimes but 32, and sometimes againe 37 or 38; the teeth are of yron, the heade of seasoned ash, and the shafte usually of saugh; betwixt the two graininges of the rake shafte they tye a stringe, which they can lappe aboute and make as longe and as shorte as they list, and then to the end of that stringe or bande they fasten a broade halters headstall, which they putte aboute theire neckes like a paire of sword-hangers, and soe trails the rake therewith.[69]

It was variously described in the nineteenth century as an heel rake, an eel rake, and could be a cumbersome 8-9 ft in width.[70] It was said by a Garsington farmer to be still in use at the turn of the century, but mostly in the hay field; 'It was always drawn behind the women with the hole in the middle of the handle so that they put their hands into the hole and drew it along behind them. It was called the "hell" rake because of the hard work in use. It kept catching in the grass you see.'[71]

Any observer, then, between 1840 and 1880, would have been compelled to remark on the crowded fields and the continuing use of hand tools. By the 1870s the mechanical reaper would still have attracted attention, but its absence would not have been remarkable on farms in the southern counties. On many a small farm, even until the turn of the century, the nails on a barn wall or the corner of some shed might still have accommodated the traditional tools of harvesting.[72] But by the 1880s all the modern aids of machine harvesting to ease the task of hand labour were in production. The self-sheaving reaper-binder, with the problem of an efficient knotting device finally solved, was available to replace or work in conjunction with the mechanical reaper, and a new diversity was observable in the harvest fields.[73] Farmers now had a choice of an innumerable variety of horse-drawn machines. The by-now commonplace mowers, 'kickers', 'tedders' and 'swathe turners' for tossing and turning the hay, horse-rakes to windrow and finally rake the fields clean, and stacking machines and elevators, working by horsepower, for both hay and sheaves, were 'all but sufficiently perfected to prove a boon to all farmers in that most laborious season'.[74]

NOTES

1 From 1563 the Statute of Labourers viz. Artificers and Apprentices Act (5 Eliz 1C4) until 1747 Regulation of Servants and Apprentices Act (20 Geo II c.19).
2 See E. E. Evans, 'Irish Folkways' (1957), p. 157; also G. Eland, 'In Bucks' (Aylesbury, 1923), p. 70. 'Sickles used by Irish gangs... more could be stored in barns.'
3 Walter Rose, 'Good Neighbours' (1942) (1949 edn.), p. 27.
4 See H. Harman, 'Buckinghamshire Dialect Glossary' (1929), p. 91; but sometimes late raking for litter was a false economy, see Hampshire and Southampton County Paper, 17 Nov. 1855, Fire in Stubble Stack, Surrounded by Wheat Bean and Other Ricks (Stanford near Farringdon, Berks).
5 Cf. 'The Nature Diaries and Note-Books of Richard Jefferies', Samuel J. Looker (ed.) (1948), p. 182. Still reaping rye in 1884 at Eltham, Surrey; note the retention of the flail for making reed, i.e. straw for thatching in North Devon c. 1890. See S. Baring Gould, 'An Old English Home' (1898), p. 209.
6 See Gertrude Jekyll, 'Old West Surrey' (1904), p. 183, 'has a square crook or step, just after it leaves the handle bringing the blade into a lower plane than the hand'.
7 But note 'strappers' as a loose Wiltshire term for any itinerant harvester in F. G. Heath, 'British Rural Life and Labour' (1911), p. 264; also the term 'foggers' for itinerant 'reapers' in late eighteenth-century Hereford in 'The Diary of a Farmer's Wife', Barbara Hargreaves (ed.) (1964), p. 50; but 'foggers' - cattle men, late nineteenth-century Oxfordshire, see local newspaper advertisements, e.g. OT, 21 Sept. 1901, Wanted at Michaelmas Married Man as 'Fogger'; see also 'vogger' in Glossary.
8 See H. E. Julyan, 'Rottingdean and the East Sussex Downs and Villages' (1948), p. 77; also 'Ag. Gaz.', 24 Aug. 1867, p. 891 for Sussex practices - 'reaping' and 'swapping' distinguished.
9 Richard Jefferies, 'Field and Hedgerow' (1948), p. 140; see also p. 132. 'The edge of the reap-hook [sic] had to be driven by force through the stout stalks.' And Ronald Blythe, 'Akenfield' (1969), p. 55. 'We reaped by hand. You could

count thirty mowers... each followed by his partner who did the sheaving.' (c. 1900 Suffolk).

10 See E. J. T. Collins, Harvest Technology and Labour Supply in Britain 1790-1870, 'Econ. Hist. Rev.', Ser. 2, vol. 22, no. 3 (1969), p. 456, fn. 1 for definitive techniques and comparative weight of tools - reap hook, 8 to 15 oz, blade length 12 to 24 in; bagging hook, 2 to 4 lb, blade length 20 to 30 in.

11 Note alternative method: 'Boil sticks in pot. Bend to shape, hang up to dry' in 'The Nature Diaries and Note-Books', p. 101; Fred Archer, 'The Distant Scene' (1967), p. 172ff, 'two hooks' (bagging hook and pick thank).

12 'Reading Farm Records', MID' 1/3/4: 'Parrotts Farm Labour Book', 27 July 1907. Hacking peas: George Sturt, 'Journals', vol. 2, p. 491 - a sickle as well as fag hook used about Farnham (29 July 1906); see Cambridge Folk Museum - small bean hook called a 'Tomahawk', also Brighton Museum - special small hook for cutting beans used at Ansty, near Cuckfield, Sussex.

13 'Ag. Gaz.', 24 Aug. 1867, p. 891. 'Corn cut with scythe... reaping-hooks for the past four or five years rarely used' (Cork).

14 See 'Year Book of Facts' (1953), pp. 114-15. Smith of Deanston in 1812 (improved model 1853), Ogle of Remington near Alnwick in 1822 (model for McCormack's c. 1826 - 2,000 annually sold in United States early 1850s), Bell of Scotland in 1829.

15 Sometimes high reaping continued into the twentieth century especially to provide cover for game (Oxfordshire), see F. G. Payne, The Retention of Simple Agricultural Techniques, 'Gwerin II' (1959), p. 130. 'I took much trouble to preserve the sickle... entirely in the interest of game', Oxfordshire farmer, 'Country Life', Oct.-Nov. 1911.

16 See 'Ag. Gaz.', 20 Aug. 1860, p. 760 for early objections: 'McConnel's Agricultural Note Book' (1883), p. 33. Quantity reaped per day 5 to 8 acres, according as one side is cut or roundabout; but see 'Bicester Herald', 10 Sep. 1875, 'Reaping machines work but slowly - they can in the majority of instances only cut one way, the return voyage being void'; note in addition, 'Some of the old makes for one horse had a very narrow cut and took a long time to get over the acres.' Primrose McConnell, 'The Complete Farmer' (1910), p. 394.

17 A labour shortage reported at Cholsey, Berks, JOJ, 21 July 1868, p. 8, col. 3. 'It is said that farm labourers were never known to be so scarce.'

18 'Ag. Gaz.', 24 Aug. 1867, p. 890, col. 2.

19 Ibid., p. 890, col. 3.

20 Ibid., p. 801, col. 1.

21 Ibid., p. 889, col. 3.

22 Ibid., p. 890, col. 2.

23 Ibid., p. 890, col. 3.

24 Ibid., p. 891, col. 2; see also 'Reading Farm Records', WIL 6/4/293, Farm Diary, Stratton, Kingston Dever (Nov. 1867). 'July 27. Put on 33 reapers extra. Began mowing oats with 13 scythes'; but 2 Aug. 1869, 'reaping machine used in cutting oats and wheat'.
25 Ag. Gaz.', 24 Aug. 1867, p. 890, col. 3.
26 Ibid., p. 889, col. 3.
27 Ibid., p. 889, col. 3.
28 Ibid., p. 891, col. 2.
29 Ibid., pp. 882-91, Harvest Work and Wages.
30 Cf. Evans, 'Irish Folkways', p. 56. 'The expert thatcher prefers his straw to be cut with a sickle and thrashed by hand and he scorns machine thrashed straw.'
31 'Windsor and Eton Express', 31 July 1875, at Langley, Bucks, sickle in use - badly laid crops; 'Reading Farm Records', BER 28/3/1, 3 women and 7 men reaping wheat at 12s per acre. One Ann Church reaped the most - Bradley Farm Estate Chievely, 1869 harvest: 'Windsor and Eton Express', 7 Aug. 1875, Dorchester (Oxon), itinerant reaper about Beaconsfield; and 21 Aug. 1875, Reaping about Datchet (Bucks).
32 See Flora Thompson, 'Larkrise to Candleford' (1965), p. 257-8.
33 OT, 17 Aug. 1889.
34 'Windsor and Eton Express', 23 Oct. 1886. In use on Colonel Follett's Estate, Windsor.
35 JOJ, 19 Aug. 1876, Reaping Trials.
36 William Potts, 'A History of Banbury' (1958), p. 224.
37 G. E. Fussell, 'Farming Techniques from Prehistoric to Modern Times' (1965), p. 198-9; between 1862 and 1892 mowers increased from 8,900 to 51,000 but see 'Implement Manufacturer's Rev.', 3 May 1875, vol. 1, p. 31 - Wood's New Iron Frame Mower, 210,613 made and sold since 1853, 20,430 made in 1874.
38 'Reading Mercury', 19 Aug. 1893.
39 Edwin Eddison, Harvesting in a Bad Season, Prize Essay, JRASE, vol. 23 (1862), p. 211, 'better to cut too soon than too late'. 'Rather let corn be muck in stook than muck in stack.'
40 JOJ, 22 Oct. 1898 (abstract of article in 'Ag. Gaz.'). But see OT 23 Aug. 1890, 'as hands are not plentiful the cutting is tedious' (S. Oxon).
41 'Implements Manufacturers Rev.', 1 Apr. 1878, vol. 3, no. 586, p. 1510 (List of Patents), Binding Corn. T. Hansen Copenhagen - improvements in apparatus for binding sheaves of corn with bands of straw or other material 12 Feb. 1878; see below p. 31. Cf. 'Reading Farm Records', HAN 7/9/1. First binder on Eversley Estate Berks-Hants border seen in 1894 (anonymous memoir Mattingley villager).
42 G. E. Moreau, 'The Departed Village' (1938), p. 77, 'see three Sundays in the field'.
43 See H. Rider Haggard, 'A Farming Year' (1899), p. 309, 'the grain is as a rule practically sapless and dead... [but] the

presence of layer being green and succulent, it takes a while to dry'.

44 'Bucks Herald', 17 Aug. 1889 (extract 'Ag. Gaz.').

45 Richard Jefferies, 'Hodge and His Masters' (1886) (1966 edn), pp. 124-5.

46 'Bucks Herald', 17 Aug. 1889, 'Time is often of more value than money in the saving of crops.'

47 E. L. Jones, 'Seasons and Prices' (1964), p. 126. Also suggests 'fewer hands were needed' but the counter suggestion: saving time was of more importance than saving on labour employed. See 'Reading Farm Records', OXF 1/1/1 f.59, 1 Oct. 1915, 'Carter Luckett away one week - no wages', Manor and Church Farms, Aston Bampton, Oxon.

48 Heath, 'British Rural Life and Labour', p. 264. 'A man and a wife will tie up and "isle" two acres a day of same, twelve to fifteen hours [reaping machine cutting]... 3/6 - 5/6 per acre' (c. 1880 Wilts).

49 See 'Ag. Gaz.', 24 Aug. 1867 and (N & Q, Ser. 8, vol. 12 (1897), p. 891. 'Haddock', 'hatrock' shock of twelve in north; see also Appendix A, p. 195.

50 Henry Best, 'Rural Economy in Yorkshire' (1641) (Publication Surtees Society, 1857), vol. 33, p. 32.

51 George Bourne, 'Lucy Bettesworth' (1913), p. 167. ' "Stand hard - Wo-o", in two notes... the second a little the higher...' (In Surrey used as a direction to the horses to remain still.)

52 John Purser, unpublished manuscript, Wellington, New Zealand, 1955, p. 23.

53 Best, 'Rural Economy in Yorkshire', p. 36.

54 Thompson, 'Larkrise', p. 255.

55 Purser, unpublished manuscript, p. 20.

56 Rose, 'Good Neighbours', p. 27.

57 Jefferies, 'Field and Hedgerow', p. 133. Cf. Bourne, 'Lucy Bettesworth', p. 109. 'In loading it is the butt of the sheaf that projects over the wagon.' Then butt to band to bind in the load - butt to band to bind the previous layer both in waggon and rick (author's note).

58 J. G. Cornish, 'Reminiscences of a Country Life' (1939), p. 81.

59 Rose, 'Good Neighbours', p. 27; cf. Thompson, 'Larkrise', p. 44. In North-East Oxfordshire in the 1880s scythes were still in use, women still reaping with the sickle; the mechanical reaper was considered an auxiliary, 'a farmer's toy', p. 257.

60 N & Q, 26 Aug. 1905, pp. 164-5. But note: the serrated sickle held its edge, if well sharpened at the beginning of harvest.

61 'Cornhill Magazine', vol. 12 (1865), p. 359.

62 Alice C. Day, 'Glimpses of Sussex Rural Life' (Kingham ND, c. 1927), pp. 20-1.

63 Mable K. Ashby, 'Joseph Ashby of Tysoe 1859-1919' (1961), p. 25.

64 Jefferies, 'Field and Hedgerow' (1948), pp. 131-2. See also Rose, 'Good Neighbours', pp. 28-9, 'sweat and sun... hardened the skin which became deeply tanned... the muscles of the arms were as knotted ropes...'.

65 C. Henry Warren, 'Adam was a Ploughman' (1942), p. 100 (from reminiscences of William Sexty, Cotswold farmer, b. 25 June 1854):

> wit to wet the scythe do cut
> the mowers be so lazy
> a pint will make 'em drunk
> and a quart will drive 'em crazy.
> (a rhyme boys sang)

66 Isaac Mead, 'Life Story of an Essex Lad' (1923), p. 25.

67 Arthur Randle, 'Sixty Years a Fenman' (1966), p. 23.

68 W. H. Barrett, 'A Fenman's Story' (1965), p. 394.

69 Best, 'Rural Economy in Yorkshire', p. 51. Note: a small land is four yards wide; an ordinary land is (at Driffield) nine or (at Elmswell) twelve yards in breadth. But see John Lubbock, 'The Beauties of Nature' (1892), p. 197, 'the most convenient unit of land for arable purposes was a furlong in length and a perch or a pole in width... the ancient ploughman... used his "pole" or "perch" [the goad 16½ ft for driving plough oxen] as a measure by placing it at right angles to his first furrow'.

70 Reading Farm Records, BUC 1/7/1, Inventory Manor Farm Horton 1862 - item 4 heel rakes, 18 small rakes; BUC 7/2/1, Inventory Hill Farm Stokenchurch 1921, scythe and heel rake; BER 21/2/1, Horfield Farm Hindred nr Wantage Sale Catalogue, 24 Sept. 1889, Lot 6, two ell rakes and 6 hay rakes - 5s.

71 Personal Communication, 11 Nov. 1969, GAR 1 Walter King, Garsington; but see Purser, unpublished manuscript, p. 23, hell rake - 6 ft wide with half-moon-shaped tines; cf. Wright, 'Old Farm Implements', pp.45-6, 'my first job... was to drag the iron-toothed rake - the "hobby" rake'.

72 Cf. Jekyll, 'Old West Surrey', p. 182.

73 See J. Hannan, Report on the Implements at the Royal Agricultural Society's Show at Liverpool and on the Trials of the Self-binding Reapers at Aigburth, JRASE, pt. 1, vol. 14, Ser. 2 (1878), pp. 131-4. With the Americans D. M. Osborne and C. H. McCormick still experimenting with wire (considered a danger to cattle), Walter A. Wood's 'nearly perfect machine' using twine was awarded a Silver Medal in recognition of 'Progress'. But by 1882, with the introduction of Appleby's knotter (1879) and abandonment of wire for twine (1881), McCormick had commenced mass producing in America. Cf. Norman E. Lee, 'Harvests and Harvesting through the Ages', (Cambridge 1960), pp. 168-71; Primrose McConnell, 'The Complete Farmer' (1910), pp. 395-7.

74 John A. Clarke, Practical Agriculture, JRASE, pt. 2, vol. 14 (1878), p. 642.

3

CROPS AND LABOUR

> The Necessities of rural life... require[d] recurrent groupings of households for common economic purposes, occasionally something like a crowd of men and women and children working together for days on end... above all [at] harvesting... We do not yet know how important this element of enforced common activity was in the life of the English rural community on the eve of industrialisation, or how much difference enclosures made in this respect. But whatever the situation was the economic transformation of the eighteenth and nineteenth centuries destroyed communality altogether in English rural life.
>
> Peter Laslett, 'The World We Have Lost' (1968), p. 12.

Traditionally, the need for labour at harvest time had always extended the resident farm labour force and even the local supply of workers. This perennial problem of obtaining adequate labour was accentuated in the middle years of the century by the great expansion in corn production.

Berkshire[1] in 1840 was chiefly an agricultural county and possessed some of the finest corn-land tracts in the kingdom, the western and central parts being the most fertile, the east and south being chiefly occupied by Windsor forest with a considerable portion of waste and unenclosed lands. The Vale of the Kennett though not as naturally fertile as the most fertile Vale of the White Horse was perhaps superior in cultivation - wheat, oats, barley, peas, beans, buck-wheat (mostly a fallow crop fed off or ploughed in), vetches, rape-seed, turnips and potatoes were extensively grown, with some cultivation of onions, carrots, hops, woad, flax, asparagus and lavender. Owing, however, to the great extent of barren heaths on the southern border and eastern part of the county, and of sheep walks on the chalk hills, the quantity of cultivated land did not at that time exceed the general county acreages, though it was this marginal land which materially increased corn production when ploughed up during the profitable years of the Crimean War and the subsequent decade and a half. As A. G. Bradley pointed out regarding West Berkshire in the early 1870s, 'grazing land was [still being] ploughed up... farmers had not yet begun to regret it'.[2]

While the amount of arable land in 1831 was computed at 255,000 acres and the natural grass land, excluding the sheep downs of the chalk districts, at 97,000 acres, in 1866 the area of permanent grass was returned at virtually the same acreage (96,264 acres) but the total acreage under cultivation stood at 451,210 acres, of which 146,844 were under corn, something approaching an additional 100,000 acres of arable.

The proportion of the population engaged in agriculture in 1831 was estimated at 45.2 per cent; in trade and manufacture 31.8 per cent. Already the rural labouring population was considered extremely numerous. Out of 31,081 families, 14,047 were engaged in agriculture and only 9,884 in trade; the agricultural labourers numbered 14,802, the factory hands a mere 321. This was understandable since earlier the county had abandoned any pretensions to figure as an important manufacturing centre and had concentrated on the introduction of new crops and better manures. By 1840, farmers were availing themselves of the products from 'several bone mills and manure manufactories' recently established - 'super phosphate of lime and guano were fertilizers used after 1841 together with extensive use of the liquid manure drill'. An interesting feature was the two-way exchange of manure and feed corn between the training stables and the downland farms at East Ilsley, Compton, Chilton Litcombe and Lambourn.[3] With the adoption of more elaborate rotations, and with the completion of the enclosure movement, the greater part of the newly fenced land was now assigned to arable cultivation.

The usual rotation of crops on the rich soils was wheat, beans, barley, oats, clover, vetches and turnips; alternatively on downland, grass (seeds) following wheat instead of beans and replacing clover, after the oats. Wheat and beans were the most productive crop in the Vale of the White Horse; wheat, barley and oats were grown everywhere, with large arable and sheep farms being the general rule, though Pearce, noting this tendency to large farms in 1794, already thought them too large.[4] At this period, 1840, artificial grasses formed an important part of the rotation of crops - rye-grass was grown chiefly in the chalk district, red or broad clover on the strong and fertile soil in general, Dutch clover, white clover, hop trefoil, sainfoin and some lucerne on the downs and uplands. The much greater commitment to corn production, however, from the 1850s until the depression, is shown by the systems of cropping in vogue in 1860. To the south and south-east of Reading a five-course rotation had been adopted: roots, barley or oats, grass, wheat, barley or oats; in the centre of the county, the common four-course: roots, barley or oats, part grass, part rape and turnips fed off for wheat; in the northern districts, roots, barley or oats, half clover, half beans, wheat; while in the vale two courses were favoured: wheat, barley, beans with a small portion of roots or wheat, beans, barley or oats, beans. Despite the effects of the depression, Berkshire farmers were to continue to favour corn production. As an example, from 1894 onwards, a sophisticated five-course rotation was worked at Park Farm, Wallingford, in conjunction with the five-field system. Twelve crops were grown every ten years, six of them corn.[5]

Buckinghamshire in 1840,[6] although it had large portions laid out in dairy and grazing farms which supplied the London market with butter, oxen, lambs, calves, hogs and early ducklings (especially from the Aylesbury district), was estimated in a contemporary report to have 352,000 acres in arable cultivation as against 170,000 in pasturage. The total estimated acreage of 518,400, however, was considered an exaggeration, a figure of 472,320 acres being a more accurate measurement, but the proportion of 2:1 arable to pasture was not queried. The estimated proportion of arable to pasture in fact roughly matched St John Priest's earlier assessment, when he stated that

> in the south part of Bucks and upon the Chiltern Hills the pasture land is very small in comparison with the arable... in the rest of the county pasture farms large dairy farms ... almost half the farms of a mixed nature.[7]

In 1831, 16,893 families were engaged in agriculture, with 8,395 engaged in handicrafts, where lace-making provided work for many women and children, though the industry was no longer so prosperous as in earlier years when Priest reported that 'no women's labour for agriculture could be obtained... owing to the good wages they were paid'. (Two decades earlier, James and Malcolm could say 'all... manufacturers together do not employ so great a number as to produce any particular effect upon the agriculture of the district'.[8]) Paper-making, straw-plait, needle-making and chair-making also provided alternative employment, occupations all virtually nonexistent in Berkshire.

Lace-making and straw-plait were carried on in Risborough and Buckland but Hanslope was the main lace-making centre where it only became extinct as late as 1884. Straw-plait was to survive the century but by 1900 only 500 were so engaged. When it was still flourishing in the 1860s, Bow Brickhill, Great Brickhill, Little Brickhill, Warendon, Aston Abbots, Drayton Parslow, Hoggeston, Pilstone, Stewkley, Swanbourne and Whitchurch, about Aylesbury and the Ivinghoe district, were the places it provided employment. Work at needle-making ended at Crendon in 1862, when the makers moved to Redditch, but the disappearance of this local employment caused only minor hardship at the time since four-fifths of the needle workers also moved. Chair-making in the Wycombe district was, in fact, the one industry that was not only to survive but to grow and where, in 1885, about 50 chairmakers large and small provided employment not only for women but also for men.

Geographically, this county fell naturally into three distinct areas - the Thames Valley, the Chiltern Hills and the Vale of Aylesbury, but both meadow and arable gave very satisfactory returns in a damp season. In a hot dry season it was noted that 'both soon burn' and on higher ground, often poor hungry flint gravel, the habit of picking stones to excess, especially in the Chilterns, did not improve conditions. Though the Vale of Aylesbury was natural grass it was rich, kindly working land and could yield fine crops of wheat and beans. Another fine stretch was about Waddesdon through Winchendon and Long Crendon, high country dividing the Thames and Ouse watershed. No strict routine of cultivation, however, was observed and the four-course Norfolk system ignored. Marshall had commented earlier on the variety of methods of tillage revealed by Priest in his report of 1813, but he did not share Priest's concern.

> The practice and opinion of nearly twenty occupiers are noted. Scarcely any two of them are alike. Yet many of them may have been 'right' under their several ATTENDANT CIRCUMSTANCES of soil and previous management.[9]

Neither did Caird who, many years later, held the view that '[t]he slavish adoption of fixed rules of rotation are suited only to a comparatively low state of farming'.[10]

Indeed, Buckinghamshire farmers continued to have minds of their own and such uniformity in cropping as evolved was based on a loose five-course rotation thus: fallow or roots, barley or oats, clover or peas, wheat, barley or oats. The lack of concern, in this method of cropping, for artificial grass indirectly indicated the extensive nature of the county's natural grass land. In fact it was ideal mixed farming country and though wheat production in the peak years between 1865 and 1875 closely followed the output of both Berkshire and Oxfordshire, permanent grass always exceeded the acreage under corn by at least 40,000 acres.[11] The county was justly famous for its joint production of corn and cattle - 'Buckinghamshire's bread and beef' was a common expression.

Oxfordshire,[12] like Buckinghamshire, had its abundance of meadows and pasturage especially in the central district. But by 1840, though extensive old pastures existing even on the fertile lands which occupied the whole of the northern part of the county were still being stringently guarded from the approach of the plough, these pastures, made fertile by the 70 streams which watered the county, were limited in extent and consequence when compared to the area which came under the plough. Unlike Berkshire, Oxfordshire was not totally bereft of significant manufacturers, but they were neither numerous nor important and in 1831 employed only 711 males and a proportionate number of females. Blankets were made at Witney, Hailey and Crawley and employed in these places respectively 200, 60 and 11 men. Plush and girth-making employed 125 at Banbury and 40 at West Shefford, Bourton and Wardington. Glove-making employed about 60 men and a large number of women at Woodstock and in its vicinity. Here too, articles of polished steel were made. The making of woollen girths and horse-cloths employed upwards of 50 persons at Chipping Norton while lace-making was still a common occupation in the southern part of the county. While tradition spoke of a considerable commerce in gloves at Bampton no 'sign of this happy state' was visible by the 1850s. The preparation of leather, too, once of some account, no longer caused foul odours from tan pits to offend the nostrils.[13] Falkner, in the last decade of the century, saw the county's fate as never to be an industrial one and, except for some tweed-making at Chipping Norton and some implement-making at Banbury, 'guiltless of manufactures'.[14]

Table 3.1: Occupations Other Than Agriculture Employing Over 1,000 Workers in 1861 and 1871 (county figures below 1,000 in parentheses)

	Berks		Bucks		Oxon	
	1861	1871	1861	1871	1861	1871
Male						
Carpenter/Joiner	2,192	2,563	1,231	1,361	1,646	1,743
Bricklayer	1,521	1,928	(831)	1,028	(367)	(522)
Shoemaker	2,047	1,647	1,882	1,686	1,606	1,321
Baker	1,213	1,762	(620)	(625)	(792)	(827)
Plumber		1,045				
Chairmaker/Cabinetmaker			1,783	2,264		
Blacksmith	1,020	1,092	(707)		(747)	
Female						
Domestic servant	6,640		3,184		5,148	
Chairmaker			(409)			
Straw-plait	(3)	(44)	2,976	3,412	91	61
				1,654*		41*
Lace	(19)	(38)	8,459	8,077	1,337	1,007
				4,442*		298*

Note: * Figures for 1881.
Source: Census Returns.

In general a contemporary opinion was that the conditions of the arable land and the arts of tillage and cropping entitled the county to a respectable rank among the agricultural districts of England. The course of crops on the lighter soils was the four-year Norfolk rotation, usually lengthened to six years with 'poulse' (a mixture of peas and beans) or oats; on the heavier soils which had been drained and lay on irretentive subsoils the whole arable land was cropped to include the sowing of artificial grass in the rotation – first turnips or other roots, next barley or oats, next three or more years of clover or grass seeds, next wheat and finally beans. When the soil was unusually heavy, beans were planted before the wheat and the land prepared in anticipation of the wheat by direct manuring or by previously feeding sheep off the herbage more than once. Corn crops were almost universally drilled and increasingly hand-hoed at the 'tiller' stage. Turnips, previously universally sown broadcast, were beginning to be generally sown in rows with intervening furrows, beans put in with a short dibble in lines three or four inches apart and a foot between the rows, while clover and rye-grass, either separately or in a mixture, was sown among the barley. Sainfoin was extensively grown on hills which were formerly matted with underwood. On chalky lands it began to thrive and yield both pasture and hay, eventually creating good arable for corn crops and either doubled or trebled the value or created 'utility out of utter barrenness'. Although the old wheel-mounted plough with a straight timber turn-furrow had been abandoned by 1840, the prodigality of horsepower and 'sluggard slowness' in ploughing still quite surprised visitors from more enterprising counties of Great Britain. Yet earlier, Arthur Young was favourably impressed by the neat round and oblong ricks of Oxfordshire, comparing them with those in East Anglia 'where the stacks resemble[d] dung hills more than they resemble[d] ricks'.[15]

Despite this evidence of a considerable formal and organised structure in agricultural practices and the establishment by 1840 of a fully-fledged commercial agriculture, contemporary descriptions relating to the early 1840s give a somewhat different perspective by a keen Scottish observer:[16]

> All round Wallingford, and from that town to Abingdon, from Abingdon to Oxford, and from Oxford to Banbury, again from Oxford to Bicester, and across the county to Buckingham, where I have recently been, everything in agriculture, whether in respect of the style of farming, the wages of labour, the condition of the labourer, or the ability and hope of the farmer, all is meagre, has ever been meagre, and is at present more than ordinarily depressed. The land, naturally good, much of it excellent, exhibits a most pitiful appearance. The miserable rents of twenty shillings an acre of land that should be paying sixty shillings, of ten shillings an acre for what should be paying thirty shillings, are not paid; the tenants cannot pay, and have no hope of being able to pay. On every acre of their land there is a demand for labour to drain, or clear weeds, or ameliorate by a mixture of soils; but there is neither the capital nor the security for capital to employ such labour.

And in the spring of 1843, in one field on a farm near Abingdon, four ploughs were at work with 24 horses drawing them, six to each plough. Indeed Berkshire farmers had a reputation for being particularly fond of their horses and 'kept too many of them'. The Commissioners in 1867 held the same opinion: 'one half of the plough boys might be dismissed and sent to school on the back of one of the horses... leaving the remaining two horses with the aid of a pair of reins and a larger allowance of corn to sharpen the step of the plough'.[17] In addition, the field ridges were serpentine, 'shapeless and measureless with furrows following their prodigal eccentricity', so that the plough travelled over an extra space of ground, as if a ten-acre field contained eleven. Numberless small enclosures and between them masses of divisional fences composed of ditches, mounds and shaggy bush were considered to be occupying space that should have been the centre of a corn field. The rural workers were seen as 'always under-fed, even if always employed' and many of them partly or wholly unemployed, 'seeking subsistence as paupers, or as poachers and thieves'.

The fields about High Wycombe were described as 'lovely despite their weedy foulness and mismanagement'. The stiff clay soil of parts of Buckinghamshire wasted the strength of six horses to a plough for the farmers knew little or nothing of the art of ameliorating clay soils. Having voted for a protective corn duty they sat down with their long pipes over their ale and the protection of that duty and lamented the hard bargain of having to pay 15s or 20s an acre for land which in Scotland would rent out at £3 10s. One opinion was that Buckinghamshire and Berkshire had 'the best land and the foulest weeds'. In front of the splendid mansion of Clivedon stretched 'the finest farms and the worst farming'. On one farm, in the harvest of 1843, the barley was absolutely wasted through the want of proper management; 'on the harvest field was... double its amount of seed, and the crops were not so luxuriant as to have much to spare'. Within a few miles of the Great Marlow Union Workhouse crowded with the unemployed poor, in the valley of the Thames, 'so rich by nature, so valueless by cultivation', a man was paring off the tops of couch and other weeds with a hoe in a bean field. 'The foulness... [so] overtopped the beans that the ground which he had hoed looked like an old pasture, or a piece of moorland newly paired of its turf.' Blenheim Palace and park were described as 'the most noble, regal like', but 'worse farming, poorer farmers and poorer labourers' were not to be found elsewhere in Britain. The estate was 'one vast wreck'. On the Earl of Abingdon's extensive estates, especially about Oxford, a farmer complained how grievously his Lordship's tenants suffered from the depredation of game. Only the previous year (1842) 'in one field only' he calculated, 'a loss of ten quarters of grain by the vermin - pheasants, hares, rabbits etc; no drawback from rent even if it was allowed could be an equivalent for such a loss'.

On the other hand, many farmers saw no point in cutting back hedges, mending fences, levelling down banks, covering in ditches, draining land, uprooting stumps and briars or planting fresh neat hedges. 'I've got no lease' was the attitude of one tenant on the Duke of Buckingham's estate, 'one year of a bad bargain is enough. The Duke can get rid of me or me of him when we tire of each other. So the place as it is will do for me; it will do my time.'

Was Somerville's vision blurred, his judgements exaggerated and harsh? In 1851, the member of Parliament for Berkshire, despite giving praise for the excellence of some farming, was sufficiently shocked at the state of agriculture to remark, 'I am quite ashamed of my county... we still continue to be disquieted in Spring by whole fields of barley as yellow as saffron from charlock in blossom...'[18] Straggling fences, untidy surfaces, gaping ditches and slow-going methods of a large farm considered well kept in the Vale of the White Horse were indeed noted by Bradley, and contrasted with the clean brisk look of the Scottish farm and their occupants. But this was in the 1870s after the Corn Laws had been forgotten, after 20 years of rising profits, increased arable cultivation and of unprecedented corn production. Yet as Bradley concluded, 'I have little doubt that the profit was as great, if not greater [on the Berkshire farm].'[19]

In 1866, the mid-point of 20 years maximum corn production throughout England and Wales, Oxfordshire ranked sixteenth with 156,338 acres and Berkshire nineteenth with 146,844 acres out of 42 counties in respect of area under corn crops. In terms of percentages of cultivated area, however, Berkshire with 42.2 per cent ranked tenth and Oxfordshire with 40.6 per cent, twelfth. Buckinghamshire with 171,192 acres of permanent grass, roughly 50,000 acres more than Oxfordshire (122,734 acres) and 75,000 more than Berkshire, had still a respectable 35.2 per cent in corn (132,691 acres), and in the several years before the subsequent decline almost matched in total production both its neighbours. Corn growing was therefore a central feature, certainly the most important in the case of Berkshire and Oxfordshire, less so in the case of Buckinghamshire, in each county's economy. Although because of their size in cultivated area they might not be considered nationally as among the greatest producers compared with counties such as Essex or Hampshire, only ten counties had a greater percentage of the male population engaged in agriculture. Bearing in mind the direct relation of overall production to handling and harvest labour requirements, the following situation emerges, demonstrating the need for harvest labour in Berkshire, Buckinghamshire and Oxfordshire. Berkshire's corn acreage remained above 1866 figures (146,844 acres) until 1878 (147,348 acres) with the really rapid decline postponed until after 1882, having by then declined by over 9,000 acres (137,430). Its wheat acreage was also still above the figure for 1866 (61,103 acres) in 1878 (61,175 acres), holding production to 1884 (53,439 acres), an overall drop of 7,664 acres before the final acceleration downwards.

Buckinghamshire's corn acreage followed very closely that of Berkshire's but at a lower level. The figures for 1878 were still above those of 1868, but over 11,000 acres down in 1882 compared with Berkshire's 9,000 acres in relation to their individual 1866 figures (Buckinghamshire, 132,691 acres in 1866, 121,242 acres in 1882); but in wheat acreage only 1,702 acres below 1866 (53,661 acres in 1866, 51,959 acres in 1882).

Oxfordshire, with approximately 10,000 more acres under corn but approximately 4,000 fewer acres in wheat than Berkshire in 1866, marginally exceeded, on average, Berkshire's wheat acreage in the three years 1872-4 and had still in 1862 a greater acreage both in corn and wheat than it had in 1865.

Table 3.2: Leading Corn-producing Counties in England and Wales, 1866 (42 counties)

	Acres		Acres
Lincolnshire	593,374	Yorkshire*	
		West Riding	220,723
Norfolk	449,432	Wiltshire	215,728
Essex	406,206	Sussex	207,793
Suffolk	405,834	Shropshire	171,177
Devonshire	271,254	Nottinghamshire	157,858
Cambridgeshire	262,597	Oxfordshire	156,338
Hampshire	260,791	Northumberland	153,716
Yorkshire*			
East Riding	259,673	Warwickshire	151,456
North Riding	251,365	Berkshire	146,844
Kent	244,494	Buckinghamshire	132,691

	% of Cultivated area
Cambridgeshire	56.7
Suffolk	52.3
Essex	51.4
Hunts	49.1
Bedfordshire	46.7
Hertfordshire	45.4
Norfolk	44.5
Lincolnshire	42.8
Yorkshire	
East Riding	42.4
Berkshire	42.2**

Notes: * Yorkshire - 3 divisions.
** Oxfordshire, 40.6 per cent; Buckinghamshire, 35.2 per cent.

Source: A & P 1866 XXIV, Miscellaneous Statistics, pp. 762-3.

Comparative figures indicate for each county the considerable persistence of corn production at a high level even into the early years of the depression.

Table 3.3: Peak Acreage in Corn Production

		Acres	% of Arable cultivation
Berkshire	1866	146,844	42.2
	1871	153,329	41.6
Buckinghamshire	1869	142,703	36.0
	1871	159,384	35.0
Oxfordshire	1869	168,239	41.7
	1875	169,508	40.9

Peak acreage in wheat production

		Acres
Berkshire	1869	64,829
	1874	65,876
Buckinghamshire	1869	60,735
	1874	60,182
Oxfordshire	1872	65,552
	1874	65,275

Source: Agricultural Returns.

Table 3.4: Some Examples of Variations in Total Area Acreage Under All Kinds of Crops and Acreage Under Corn Crops for 1874 (nine counties of southern England)

County	Total area in statute acres	Total acreage under all kinds of crops	Corn crops
Berks	450,132	366,908	149,804
Bucks	467,009	399,168	136,419
Cambridgeshire	524,926	482,634	257,537
Devonshire	1,655,161	1,076,223	294,951
Essex	1,055,133	818,865	418,757
Gloucestershire	804,977	642,835	176,402
Hants	1,032,105	696,768	261,815
Kent	1,004,984	724,548	249,294
Oxon	470,095	413,827	167,140

Source: Agricultural Returns.

Table 3.5: Comparative Figures for Eight Corn-growing Counties in Southern England (average acreage in wheat during four periods in hundreds of acres, percentage of cultivated area in wheat in parentheses)

	10 years 1872-81	5 years 1882-6	5 years 1887-91	3 years 1892-4
Beds	504 (19.5)	448	446 (17.2)	391 (16.9)
Berks	594 (16.0)	514	469 (12.5)	407 (10.9)
Bucks	546 (13.6)	469	440 (11.0)	379 (9.4)
Cambs	1,251 (25.9)	1,097	1,087 (22.2)	1,016 (20.8)
Hunts	449 (21.5)	393	374 (17.7)	330 (15.6)
Lincs	2,818 (19.1)	2,390	2,306 (15.2)	1,904 (12.5)
Oxon	599 (14.1)	514	465 (11.2)	387 (9.3)
Warks	686 (14.1)	519	463 (9.3)	386 (7.8)

Source: PP 1894 (C7400-111) XVI, RC on Agricultural Depression, App. D, p. 619, Wheat Area Table VIII (extract).

Table 3.6: Wheat Acreage and Yields in England, 1888-98

1888	2,418,674 (28.18)*	1894	1,826,626 (30.71)
1889	2,321,504 (29.87)	1895	1,339,806 (26.21)
1890	2,555,694 (30.79)	1896	1,609,255 (33.88)
1891	2,192,393 (31.33)	1897	1,785,562 (28.97)
1892	2,102,969 (26.20)	1898	1,987,385 (34.76)*
1893	1,798,869 (25.81)		

Note: * Yield approximately one million bushels more in 1898 than in 1888.

Source: A & P 1899 CVI, Agricultural Statistics, p. 143.

As late as 1884 the respective acreages under the most labour intensive crop i.e. wheat, were as follows:

	Wheat	Acreage under corn crops	Acreage under green crops	Total
Berkshire	53,439 (29.9%)	132,235	55,282	187,517
Buckinghamshire	49,003 (32.7%)	117,070	33,406	150,476
Oxfordshire	53,583 (26.5%)	149,046	53,643	202,689[20]

These figures can be compared with the national acreage of wheat in Great Britain which amounted to:

Year	Acres	% of Arable acreage
1868	3,652,125	20
1870	3,500,543	19
1880	2,909,438	16.4
1890	2,386,336	14.2
1900	1,845,042	14.2
1910	1,808,854	12.3

Table 3.7 gives numbers of agricultural labourers in Berkshire, Buckinghamshire and Oxfordshire between 1841 and 1901.

Table 3.7: Numbers of Agricultural Labourers (male and female shepherds and indoor servants)

Year	Berks	Bucks	Oxon
1841	18,469	18,697	17,727
Farmers	1,880 (10)[1]	2,471 (8)	2,365 (8)
1851[2]	30,347	20,697	24,650
Farmers	2,166 (14)	2,033 (10)	2,468 (10)
1861[3]	28,745	19,745	25,156
Farmers	1,985 (14)	2,010 (10)	2,265 (11)
1871	25,978	19,525	22,421
Farmers	2,164 (12)	1,839 (10)	2,298 (10)
1881	20,780	13,862	18,574
Farmers	1,621 (13)	1,717 (8)	1,782 (10)
1891	17,655	12,781	15,839
Farmers	1,586 (11)	1,836 (7)	1,851 (9)
1901[4]	10,526	10,959	11,474
Farmers	1,254 (8)	2,145 (5)	1,796 (7)

Notes: [1] Approximate number of labourers to each farm (rounded up).

[2] Approximate number of labourers to each farm: Berkshire, 14; Buckinghamshire, 10; Oxfordshire, 10. (Compare Bedfordshire, 13; Cambridgeshire, 8; Northumberland, 6; Nottinghamshire, 5; Rutland, 4; Cornwall 3.5; Cheshire, 3.)

[3] Only 10 counties had a higher proportion in rural occupations than Oxfordshire: Oxfordshire 27.1% (11th); Berkshire 25.6% (14th); Buckinghamshire 25.4% (15th).

[4]

	Shepherds	Cattle	Horses
Berkshire			
Hired men 5,102:	654	1,543	2,962
day labourers 5,213			
Buckinghamshire			
Hired men 4,851:	536	2,010	2,?38
day labourers 6,048			
Oxfordshire			
Hired men 5,377:	951	1,560	2,891
day labourers 6,019			

Compare number of shepherds:	1871	1881
Berkshire	538	980
Buckinghamshire	262	394
Oxfordshire	339	823

Source: Census Returns (Occupational Abstracts).

Although a switch in hand-tool usage and gradual mechanisation was one factor that made increased production possible, as has been shown in Chapter 2, technological progress alone does not adequately explain how Oxfordshire, for example, achieved greater production with lower numbers of agricultural workers than Berkshire, or how Buckinghamshire with even lower numbers, with home industry competing for labour and a greater emphasis on mixed farming reached such a comparatively high production in corn. Available labour and its organisation must be examined for a more complete explanation.

Unlike the situation in many corn-growing counties, a significant feature existing in the three counties under discussion was an indigenous population which could supply considerable extra labour at busy periods in the farming calendar. This pool of labour resided in the open villages and might be described as the core of harvest labour requirements. It was not so much the numbers of day labourers but the help provided within the family circle that made the vital difference. Extracts from a contemporary report illustrate the kind of family help available:

> 'girls and boys under 10 years of age are employed by their parents in harvest time' (Asthall, Oxon); 'girls... help parents reap and bind the sheaves' (Stokenchurch); 'Girls of all ages are employed on harvest... with their mothers' (Swyncombe); 'I help him with a little hay, we can cut and tie an acre a day, and we got 9s an acre last harvest' (Labourer's wife); 'Women are employed... at piece work with their husbands hoeing, reaping and forking and raking corn' (Mr Luke Lowsley); 'labourers' wives generally... work the greater part of the year' (Mr Kingham Reaves); 'very generally employed... haymaking, reaping' (Henry Tucker) - situation in Berkshire.[21]

Accordingly, the day labourers with this family help could accelerate the speed with which the first stage of the harvesting process was undertaken - cutting, tying and shocking the corn - while the method of payment by piece work encouraged them to telescope the period engaged in these tasks by extending, through very early and late working, the harvest day. In no sense however is this a complete answer for there was no simple balance between local farm needs at harvest time and locally available labour; the largest corn-growing farms were not necessarily sited close to the most populous villages. While agricultural workers themselves, therefore, did not represent the total harvest labour force, the main problem was one of deployment rather than of any absolute shortage. It is true that pockets of absolute shortage did occur, but these arose mainly because of the impossibility of any exact calculation of labour requirements until harvest time itself. How then did the labour situation resolve itself in practice?

For the annually hired labourers the extra summer earnings, apart from the Michaelmas Money paid at the end of their year's service, were restricted to the harvesting of the hay and corn crops on their master's farm. This opportunity of adding to the weekly wage (usually 1s a week more than day men in consideration of their Sunday work) was confined to about a fortnight's reaping or fagging and an increased wage of 4d a day for another fortnight when leading corn.[22] Day labourers living in the open villages, free to remain so long as the rent was paid, were also free to take whatever employment there was, near or far. These villages were used to movement even in winter. Dr Hunter reported in 1865 that 'a crowd of country people had lately come to live [at Bicester] but not to work in the town'. 'People live here [Eynsham] who have to walk three miles to their work. Stanton Harcourt is said to be one of those closed villages which take the labour of the poor but decline to give them shelter, or to take the burden of their sickness or old age... [A Whitney] old man walks daily to and from his work at Braze Norton, four miles off.' Inhabitants of Abingdon worked in Drayton parish; 'much of the extra labour about Shiplake came from Reading in the sixties' as 'cots had been destroyed and people who worked so much as four miles off were driven to seek lodgings [there]'.[23] Those who lived in Bladon and in Hessington and in Coombe on the perimeter of Blenheim Park, and those living in Wootton further to the north, worked on the Blenheim Estate either regularly or intermittently in the winter months.[24] In summer there was always harvest work 'somewhereabouts'. In 1851, of 52 labourers residing in Neithrop only 25 were employed by farmers within Neithrop itself; 'it seems that the township housed workers on farms from all over the western side of Banbury parish... many Neithrop agriculturists were casual labourers'.[25] The villagers of Wolvercote above Oxford crossed the river to take in the harvest in Wytham, an enclosed parish in Berkshire;[26] the villagers of Dorchester below Oxford crossed the river to cock the clover hay in the meadows above Wittenham.[27] Men of the open village of Wheatley went harvesting around the district. In the corn harvest of 1892 a belligerent Wheatley labourer assaulted an old labourer in a field at Chilworth in a dispute about sheaves - they both had had too much to drink.[28] A Brill band of harvesters abandoned a Chilton farmer's crops in 1869 when, having agreed to take the crops 'rough and smooth' at 12s per acre, other workers commenced cutting the lighter pieces.[29]

The distances travelled in all these instances were comparatively small, within a radius of, say, less than five miles. But mobility was not just confined to reasonable walking distance when there was harvest work to be had, for this extra summer earning was vital to offset winter's uncertainties. Jefferies states that in the 1870s '[T]he young labourer... is rarely tied by the year... he prefers to be free so that when harvest comes he may go where wages chance to be the highest.' He elaborates further: 'the young labourer... has become somewhat of a wanderer. He wanders about, not only from village to village but from county to county.'[30] Collins makes it clear how widespread and established the practice was more than 20 years earlier:

> By the 1850s it had in some areas become almost a tradition that ambitious young men should take to the roads in the

> summer, selling their labour to the highest bidder. In others the flow was more regularised. North Hampshire men went into Sussex and Wiltshire, Stour Valley scythesmen into Foulness and the Marsh Hundreds, Vale of Taunton men on to the Mendips, Vale of White Horse men on to the Wiltshire Downs, Blackmoor Vale men on to the Dorset Downs, Dorset and Somerset men into the Isle of Wight, Dalesmen into the Yorkshire Wolds, Farnham baggers down to Brighton, and Aberdeen mowers to the Lowlands and Border Counties.[31]

It is not surprising then that in the 1860s men came to the Eton district for harvest work from North Buckinghamshire, Berkshire and even Wiltshire.[32] Woodstock men were to be found 20 miles south of Oxford, though only to be assaulted in 1897 by three Ipsden men in the case of a Woodstock harvester because he was deemed a 'stranger'.[33] Buckinghamshire men were there in harvest fields about Banbury 20 miles away in the summer of 1860.[34] Bladon villagers travelled a similar distance to Ballscot, four miles north of Banbury, where in 1873 three of their number assaulted three Irish harvesters, driving them out of the harvest fields - back to Ireland.[35] Certain 'open' villagers formed themselves into 'bands' as harvest time approached. Joseph Arch, the fortunate possessor of an inherited cottage at Barford in South Warwickshire, 'ever bent on improving [his] condition and earning a better wage', recalls in one of the few rural workers' autobiographies of the nineteenth century how his own 'company' was formed in the 1840s:[36]

> I went into different English counties, and also into Wales, hedge-cutting. I got good jobs, and very good money, and was in great request. Not only was I a master of this branch of my craft, with men working under me, but as I had taken to mowing when sixteen years of age, I had now become a master hand at that also, and had almost invariably a gang of from twenty to twenty-five men under me in the field... I made very good mowing contracts with large graziers: they would give me six and seven shillings an acre. The farmers were not so liberal by half, as they seldom paid more than three shillings an acre.[37]

These travelling 'bands' created for themselves their own 'harvest circuits'. They moved away from their cottage homes in the early days of May, to mow early hay in distant places, and trod the same elliptic path year after year, returning through cornfields which they reaped, and in the later years of this century 'fagged' or scythed as they went, until they returned home to harvest the corn in their own neighbourhood.

The inhabitants of Filkins in East Oxfordshire were just such travelling harvesters:

> They used to go to London mowing the parks and fields with the scythe; then helped to make the hay and put it into ricks. When the haymaking was done they worked their way back by doing hoeing for market gardeners. After working a week in one place, they walked on Sunday a few miles nearer home, and by the time they got to Wantage, the harvesting was ready as it was earlier on the downs. After cutting the corn with their fagging hooks, the harvest at Filkins itself was ready; then they would go on to Northleach where the harvest was later still.[38]

The last journey to these hayfields was made in 1912. The villagers of Bampton and Chipping Warden above Banbury just across the county boundary in Northamptonshire, and of those hamlets between, Minster Lovell and Fields Assaerts, Leafield and Spelsbury, who were transmogrified into morris dancers whenever time and tiredness allowed, would some of them go to London and the south, also to work at this early hay harvest, and tramp northwards for the later harvest. During their tour they would give exhibitions of dancing and singing, and strange villages would hear the songs they sang, 'Green Garters', 'The Maid o' the Mill', 'Trunk Horse', 'White Sock', 'Moll o' the Whad', perhaps for the first time. And in this way they would increase their harvest wages.[39] Edwin Turner of Finstock was one of these travelling harvesters in the 1860s:

> He and others used to go up to London every summer for haytime and would often go up a month sooner than necessary in order to morris in the streets, would go out day by day in different parts of London - Clerkenwell was the only name he could remember - often made 10 or 11 shilling a piece per diem. Once a fiddler from Churchill ('We tak him with us; thought it wd do him good') stole the takings.[40]

Edwin Turner indeed went dancing and haymaking 19 consecutive summers; walked there and back. Individuals from Asthall and from Buckingham went taking their 'Pipe and Tabor', and their whittle and club, and sometimes their wives.[41] There would be all these harvesters from the eastern part of Oxfordshire meeting in the hay fields of Middlesex.

Harvesters, too, from other counties would be there such as the Maddenham villagers from Buckinghamshire a few miles from Thame who, as an established custom continued into the 1870s, used to travel 'uppards' in Maytime:

> Each hay season... wandering labourers flocked there from London and the country around. Men of our village went regularly and always found employment at once; for the men of Bucks were known, and preferred to those from the town. They started out early on a Monday morning, men in groups of three or four who had agreed together to take mowing by

> the piece, in co-operation. They knew one another's capabilities, each would be able to maintain the pace, and at the same time to cut a swathe of the normal width... They called this migration going uppards. If anyone asked for them... the reply would be 'Gone uppards', for everyone was sure to know what that meant.
>
> They carried their scythes with them; blades of proven steel, ground ready, wrapped in old sacking, and carried separate from the curved sneds and wedges, and a hammer to tighten them, were taken... in a basket, where they also stored the whetstone, or rubber. Each knew that the scythe was the essential co-operator; they knew its 'right hang' to a nicety, the angle at which it would cut the heaviest swathe with the least effort...
>
> They slept on shakedowns of straw or hay in the outbuildings of the farmsteads. Beer was supplied free and in plenty...[42]

Jack Lenten, a Warwickshire thatcher, the 'cleverest labourer in the village', was another who went towards London whenever the hay-making season came on, but he found that he could not get regular employment when he returned. The farmers told him 'he might go in winter where he had been in summer':

> He had to get another place but they would never let Jack gain a settlement, so in bad times Jack was sent to his own parish to find that his cottage was pulled down. He had to live miles away and tear his heart's blood out walking to and from his work night and morn... When the railways came [c. 1850] Jack got work as a ganger...[43]

A Peter Paistow was also unlucky. He 'ran away on the 15 June and went uppards to hay work' but on his return was charged in late August 'with absconding himself from service'.[44]

The need for labour at harvest time was never precise; each village, each farm, had its own saturation point. But the state of the corn, whether upstanding or badly flattened by a past or immediate storm, and the concurrent ripening of crops made exact estimates of labour requirements a matter of uncertainty.[45] The farmer was not concerned who took his harvest so long as it was safely gathered. Occupational barriers therefore could disappear at harvest time. Because harvest was the central operation of the farming year - because no two harvests were alike - getting the harvest was a time of involvement, often of anxiety, sometimes of crisis. Before the horse-drawn reaping machine had appeared in the early 1860s and 1870s the growing problem of hand labour had become so pressing that:

> the carpenters and wheelwrights left their benches, the masons laid aside their trowels and all others left their crafts to get the precious corn in. If they had not gone the farmers

> would have withheld their patronage during the ensuing winter.[46]

In August 1872 two building labourers demanded 7s 6d a day to dig a site for the new gasholder at Woodstock, because the harvest was about to begin. They were refused this 'exhorbitant' rate and so went harvesting.[47] Earlier, in the harvest at Sutton, in building a rick there were employed, and simultaneously, the following professions: 'viz. one gentleman, two carpenters, two shoemakers, one wheelwright, one collar maker, one miller, one labourer and one land surveyor... strange to say the rick [stood] a memorial to the builders' skill and perseverance'.[48]

A good deal of casual labour came out of the market towns, 'men and women, and girls glad of the open air work'.[49] But a special kind of casual help was supplied by the villagers of Headington Quarry, an independent 'open' village with a 'closed' antagonism to strangers, the inhabitants an amalgam of quarrymen and gypsies finding a precarious living from alternate urban and rural employment. The following verbal communications relating to the turn of the century may not be altogether representative of rural Oxfordshire but they well illustrate the complexities of harvest time, the extra earnings that could be picked up at some favourable moment, how precarious economic survival was for the rural poor in the nineteenth century and sometimes how fortuitous. Bert Gurl recalled of his father:

> In the summer time... they'd go harvesting fagging and that. Mother used to go along with us. I can remember making the bands for them when I was a toddler. He went all over the place... that's cutting the corn, shocking it up - so much an acre they used to take it on... piecework.[50]

He talked about Mark Cox, his uncle, who had been in the Guards, had done some work labouring, might have done a bit of claymaking, but work, 'not a lot of anything': 'But he'd go out with the scythe in the spring of the year and do a bit of mowing; he was one of the best with the scythe...'[51]

Charles 'Waggle' Ward, now in his seventies, was a thatcher and hay-tier who used to thatch the ricks in the Quarry farmyards. In the harvest he used his skill hay-tieing and thatching - 'same job same material' - to tie sheaves at harvest time. Charlie Cooper described how he worked:

> he used to whip these long cords of straw, it was just twisted in his hand - it used to come out like magic, like a skein of wool, and he whipped that round a bundle, and give it a couple of twists, took it back in, and sling it aside... it was just magic to me then.[52]

In the winter Charlie Ward and his father and uncles went thatching houses and ricks, and according to Bill Webber, a fellow schoolboy in the 1900s:

> they used to have the pony and they'd drive out to Gorrington, Stadhampton, out to Chalgrove anywhere, hay tieing and straw tieing... that was really their job... these hay merchants used to buy ricks, and [they] used to go and tie 'em [for them].[53]

In the summer a farmer would have a field of hay - 'they used to put it into ricks'.[54] Bert Gurl recalled how Bobby Cooper used to go down to Bayswater Farm, the other side of the main London Road: 'He used to go there and do quite a bit - harvest'n hay or whatever was going - as a spare man if it was a rushed job...' He gave further details of the time when his father went cutting the corn:

> [He] used to take it on by th'acre... Fagging, that's the only way they called it. He'd go round several farms, where-ever there was anything going, to keep a crust of bread in the house. He'd go and do it, when they was out of work...[55]

Will Webb spoke about an old chap who told him how they went out mowing:

> they used to get up at three o'clock in the morning and go out on the Clayhills - two or three of them - they went there with the scythe... it's sweated labour... they used to... take plenty of beer with 'em - they used to take it on for so much, to do the lot.[56]

Walter Trafford related how he went hay making for a George Lord 'when Green Road and all that was fields': 'I went up there hay making for two or three days, then I went to Beckley... [helped] ... with some corn they was a-threshing, [then]... went across to Beckley'.[57] Charlie Cooper explained why there was so much casual work available at this period:

> These farmers - well smallholders a lot of them were, but I mean they called them farmers - if they got a crop, just been cut say, and it [the weather] didn't look too promising, they'd come round the village and rope anybody in - don't matter, women, anyone, to go and get... the waggons into [the] barn, that was the main idea, or in the rick... as long as they got it in the rickyard they felt safe... I've gone many a time and helped chuck the shocks on to the waggon... [at] Lord's farm or Church farm...[58]

These quarrymen were just as much part of the harvest labour force as migrant labour or the Irish harvesters, or those 'own' men on a Garsington farm in 1904 who worked alongside the farmer and his son without any outside help or for that matter extra harvest payment. They worked in the harvest fields from dawn to dusk for the current rate of 14s a week with free beer as the only extra. But often they got so drunk and so quarrelsome that no further work was done after eight o'clock in the evening.[59] But the true casual workers were the occasional off-duty soldiers who helped at harvest time in the vicinity of their barracks, or the gypsies who sometimes but not often would lend a hand - though with the Tadley gypsies on the Berkshire-Hampshire border it was a tradition which lasted into the twentieth century. Finally, of course, there were the tramps - one such casual worker one summer in the 1840s tramped from London because there was no work there, to find harvest work about Oxford, earning enough in two months to keep himself and his wife at Moreton-in-Marsh through the winter without further work.[60] Another in the 1870s, an elderly resident of Shepherds Bush, London, 'set out on tramp every June, worked in the Middlesex hay fields, harvested in Hertfordshire, and working his way at any kind of field work, finished up with the northern late corn harvest'.[61] But 'as for "tramp" labour meaning the true "tramp" or "cadger"', Jefferies thought it 'quite valueless, and simply a nuisance'. 'A "cadger" will work about two days, get a few shillings advanced on him on some pretext, and then decamp with two or three small articles - whatever he can lay hands on.'[62]

In addition, however, harvest work provided extra earnings to others who were not directly involved. On larger farms where piece work was extensive, settling payment was not always a simple matter. The extra money earned at piece work was often allowed to 'run on' and full settlement made at the end of the harvest or at Michaelmas - often there was difficulty in proving a claim to piece work payment when a job was finished.[63] To protect both the interests of the worker and the farmer, 'measurers' were sometimes employed. Harvesters cut the corn in drifts down the whole length of the field usually following the ploughed 'lands', the strips of uniform width parted by water furrows. But the drifts could vary in size and if they were to be properly measured definite marks were needed on the stubble. In Sussex these drifts were called 'cants'.[64] In Surrey[65] and Buckinghamshire[66] the essential reapers' 'marks' were made by tying together in a loose knot half a dozen straws of rooted corn with the ears removed. The measurer was given a helper who carried the long chain with which the drifts were measured, and who provided the names of the workers who cut them. The information was then given to the farmer who paid him his fees, but half of the amount was deducted from the workers' earnings. Deducting the whole amount appeared to be the practice on the Ditchley Estate, certainly in 1863. The Account Book has the following entries:

> Edward Harling & Co. On Acct. Mowing - £5 (Aug 21); mowing barley 11a 2r op at 2/6 etc. Total £8.4.3 - less measuring 6/3, less £5 received - paid £2.18/- (Sept 18).[67]

But the following two of 12 separate entries from the Labour Book of Fawcett Farm, Oxon, indicate that the 'lands' could be used as a guide to acreage without employing a paid measurer:

> 9 lands - 2a 3r 14p. Wheat @ 11/- - £1.11.3.
> 4 lands - 1a 3r 38p. Wheat @ 12/- - £1.4.0.[68]

The measurements were surprisingly meticulous, as detailed in the labour books of the period, and equally exact were the payments which were often calculated to the last farthing.[69] Sometimes the measurer was a professional taking any form of measuring, reaping, hoeing or thatching (by the 'square' for thatch); often he was anyone in the village who had learned to read, write and to cast up figures; sometimes he might be a carpenter, sometimes even a schoolmaster.

When, therefore, the many types of worker who were responsible for eventually gathering in the harvest even late in the nineteenth century are considered in detail, those described as present in the fields on Tom Strong's Stubble Farm in Berkshire do not seem such an improbable collection of harvesters:

> all sizes and ages, men and women, Irish and English, strollers and neighbours, reapers and faggers, good workmen and bad, grandmother and child, kettle boiler and tier, married and single.[70]

All reserves of labour were drawn upon to meet increased corn production; it can be inferred that the 18 women who were haymaking at 1s per day on an Amersham farm in 1866, or the practice of employing boys from eight years old - 'some all the year round... others in the spring and the harvest...'[71] - or the Headington Quarry boy who 'shocked up' in corn harvest at the turn of the century and complained 'I had nothing for my day's work',[72] or even the aged woman of 84 who helped her son when wheat cutting on Ravenscott Farm, Pamper Green, Berkshire, in the harvest of 1893,[73] represented the rule rather than the exception. The only certainty is that it made less uncertain the labourers' struggle to support their families.

NOTES

1 Descriptive passages based on the following sources: Samuel Lewis, 'Topographical Dictionary of England', 2nd edn (1833), vol. 2, Berkshire; 'Parliamentary Gazetteer of England and Wales 1840-1' (Glasgow, 1842), pp. 163-4; 'Victoria County History of Berkshire' (1906), vol. 1, pp. 371-412; 'Social and Economic History' (1907), vol. 1, pp. 167-243, Industry; J. B. Spearing, On the Agriculture of Berkshire, Prize Essay, JRASE, vol. 21 (1860); J. C. Clutterbuck, The Agriculture of Berks, JRASE (1861), vol. 22. See also Berkshire Census Returns (percentage increase in parentheses): 1841 - 161,759 (10.6%); 1851 - 170,065 (5.1%); 1861 - 176,256 (3.6%); 1871 -

196,475 (11.5%); 1881 - 218,363 (11.1%); 1891 - 239,138 (9.5%); 1901 - 256,509 (7.3%).

2 A. G. Bradley, 'When Squires and Farmers Thrived' (1927), p. 193.

3 Cf. 'Victoria History', vol. 2, p. 335.

4 W. Pearce, 'General View of the Agriculture of Berkshire' (1794), p. 19.

5 Reading Farm Records, BER 19/7/1.

6 Descriptive passages based on the following sources: Lewis, 'Topographical Dictionary', vol. 2, Buckinghamshire; 'Parliamentary Gazetteer', pp. 305-7; 'Victoria County History of Buckinghamshire' (1905), vol. 1, pp. 397-402, Agriculture, and (1908), vol. 2, pp. 37-95; 'Social and Economic History', pp. 103-29, Industry; Clare Sewell-Read, On the Farming of Buckinghamshire, Prize Essay, JRASE, vol. 16 (1855). See also Buckinghamshire Census Returns (percentage increase in parentheses): 1841 - 156,439 (6.4%); 1851 - 163,723 (4.7%); 1861 - 167,993 (2.6%); 1871 - 175,926 (4.7%); 1881 - 176,155 (0.1%); 1891 - 185,284 (5.2%); 1901 - 195,764 (5.7%).

7 St John Priest, 'General View ... Buckinghamshire' (1813).

8 But see James & Malcolm, 'General View ... Buckinghamshire' (1794), p. 46.

9 W. Marshall, 'Review and Abstracts of the County Reports (Board of Agriculture)', vol. IV, Midland Department (1817), p. 535.

10 J. Caird, High Farming Vindicated and Further Illustrated, 2nd edn (1850), Pamphlet Collection, vol. 26, Oxford Union Library, p. 17.

11 A & P 1866 LXXIV, 'Miscellaneous Statistics', p. 762. Peak corn production 1866 - 132,691 (35.2%) acres - permanent grass 171,192 acres.

12 Descriptive passages based on the following sources: Lewis, 'Topographical Dictionary', vol. 3, Oxforeshire; 'Parliamentary Gazetteer', pt. 9, pp. 576-80; 'Victoria County History of Oxfordshire' (1907), vol. 2, pp. 255-7, Industry, and pp. 279-92, Agriculture; R. Davis (of Lewknor) 'General View of the Agriculture of Oxfordshire' (1794); A. Young, 'General View of the Agriculture of Oxfordshire' (1809); Clare Sewell-Read, On the Farming of Oxfordshire, Prize Essay, JRASE, vol. 15 (1854). See also Oxfordshire Census returns (percentage increase in parentheses): 1841 - 163,127 (6.3%); 1851 - 170,439 (4.5%); 1861 - 170,944 (0.3%); 1871 - 177,928 (4.1%); 1881 - 179,559 (0.9%); 1891 - 185,240 (3.2%); 1901 - 181,120 (2.2%).

13 'History Gazetteer and Directory of ... Oxford' (Peterborough, 1852), p. 476.

14 J. M. Falkner, 'A History of Oxfordshire' (1899), p. 304.

15 A. Young, 'Farmer's Calendar' (1813 edn), p. 151.

16 See A. Somerville, 'The Whistler at the Plough' (Manchester, 1852), pp. 21-3, 102-5, 126-9, 140-1, for this quote and subsequent quotes in text.

17 Cf. PP 1868-9 (4202) XIII, RC on the Employment of Children, Young Persons and Women (1867), pp. 155-79; use of horses in Berkshire and Oxfordshire (44).
18 Thomas Pusey, 'What Ought Landlords and Farmers To do?' (1851), Collection of Pamphlets, vol. 26, Oxford Union Library, p. 44.
19 Bradley, 'When Squires and Farmers Thrived', p. 183.
20 For corn and wheat acreages for all three counties, see Table 3.3, p. 43.
21 PP 1868-9 (4202-II) XIII, RC Employment in Agriculture, pp. 329, 330, 341 and 371.
22 See PP 1868-9 (4202) XIII, p. 150.
23 PP 1865 (3484) XXVI, Inquiry on the State of Dwellings of Rural Labourers by Dr Hunter, pp. 254, 253, 252, 155, 153.
24 PP 1868-9 (4202-II) XIII, pp. 570, 571, 575.
25 Barry S. Trinder, Banbury's Poor in 1850, 'Banbury Historical Society Pamphlet' (1966), pp. 108-9.
26 PP 1868-9 (4202-II) XIII, p. 575.
27 OT, 30 Aug. 1890.
28 Ibid., 17 Sept. 1892.
29 JOJ, 17 Sept. 1859.
30 Richard Jefferies, 'Hodge and His Masters' (1886) (1966 edn), vol. II, pp. 51 and 67.
31 E. J. T. Collins, Harvest Technology and Labour Supply in Britain, 1790-1870, 'Econ. Hist. Rev.', 2nd series, vol. XXII, no. 3 (1969), pp. 470-1.
32 PP 1867-68 (4068-I) XVII, RC on Employment, App. pt II, p. 538.
33 OT, 13 Aug. 1897.
34 JOJ, 8 Sept. 1860.
35 Ibid., 27 Sept. 1873.
36 E.g. Alexander Somerville, 'The Autobiography of a Working Man', Hon. Eleanor Eden (ed.) (1862); William of Milne, 'Reminiscences of an Old Boy 1832 to 1856' (Forfar, 1901); George Edwards, 'From Crow Scaring to Westminster' (1922).
37 Joseph Arch, 'The Story of His Life' (1898), pp. 39-40.
38 Bodleian MS Top Oxon c220, Notes of Thomas Banting of Filkins, 1887, ff.65r 96r (abridged).
39 Percy Manning, Some Oxfordshire Seasonal Festivals, 'Folklore' (1897), vol. 8, pp. 318-19.
40 Clare College Library, Cambridge, Cecil Sharp MSS, Folk Dance, Notes II, pp. 129-30.
41 Ibid., pp. 95 and 100.
42 Walter Rose, 'Good Neighbours' (1942) (1949 edn), pp. 79-80, also 'Fifty Years Ago' (1920), pp. 26-7.
43 J. T. Burgess, 'Life and Experience of a Warwickshire Labourer' (1872), quoted in A. W. Ashby, 'One Hundred Years of Poor Law Administration', vol. III, Oxford Studies in Social and Legal History, Paul Vinogradoff (ed.), Oxford 1912), pp. 79-80.
44 'The Aylesbury News', 29 Aug. 1840.

45 Cf. E. L. Jones, 'Seasons and Prices' (London, 1964), Ch. V.
46 Rose, 'Good Neighbours', p. 27.
47 JOJ, 3 Aug. 1872.
48 'Banbury Guardian', 14 Sept. 1848.
49 Richard Jefferies, 'Chronicles of the Hedges', Samuel J. Looker (ed.) (1948 edn), p. 184.
50 Private communication, Ruskin College, Oxford - Gurl/8.
51 Ibid., Gurl/18.
52 Ibid., Cooper/C.
53 Ibid., Webber d/7-9.
54 Ibid., Cooper/C.
55 Ibid., Gurl/9.
56 Ibid., Webb d/3.
57 Ibid., Trafford 2/3.
58 Ibid., Cooper C/6.
59 Personal communication, GAR 1.
60 Somerville, 'Autobiography of a Working Man', pp. 44-5.
61 James Greenwood, 'On the Tramp' (1885), p. 26.
62 Jefferies, 'Chronicles of the Hedges', p. 184.
63 PP 1893-4 (c 6894-II) XXXV, RC ... The Agricultural Labourer, App. BII Wantage, pp. 220-1. Complaints at East and West Hannay - men lose money.
64 See Alice C. Day, 'Glimpses of Sussex Rural Life' (c. 1927).
65 George Sturt, 'The Journals of George Sturt', Geoffrey Grigson (ed.) (1941), pp. 439-40.
66 Rose, 'Good Neighbours', pp. 28-9.
67 Oxford RO, DIL 1/e/2c Ditchley Est. Farm a/c Book, 1863.
68 Reading Farm Records, OXF 2/2/4, Fawcett Farm, Oxon Labour Book, p. 22.
69 Ibid., BER 28/3/1, Bradley Farm Labour Book (1869): 'Thos. Church oat fagging 1a. Or 31p. at 3/- per acre. 3/6$\frac{3}{4}$. Wm. Holt 2a 1r 17p. 7s.0$\frac{3}{4}$d', and 5 other entries.
70 H. Simmons, 'Stubble Farm' (1880), vol. II, pp. 17-18.
71 PP 1867-8 (4068-I) XVII, RC Employment ... in Agriculture (1867), App. II, pp. 535, 539.
72 Private communication, Gurl/9.
73 'Reading Mercury', 12 Aug. 1893.

4

CHILDREN'S LABOUR AND EDUCATION

Juvenile Labour.

Salisbury Plain knows little about summer holidays. The children are out of school but they are helping to get in the harvest as they have been helping to get in the hay. All day long the farm children are out playing or working on the downsides in the hot scented air, the broad cornlands spread away to the south where from the summit of every undulation can be seen, rising from golden hillsides, a dark spike, which is Salisbury Spire.

'Times Education Supplement', 19 Aug. 1920

Until 1870 children, especially boys, were extensively employed in agriculture at an early age; sometimes only casually, often when growing older as annually hired hands, virtually always in some capacity in the harvest fields. Education as a restricting element on their employment was minimal as the following figures seem to suggest. In 1849 in Oxfordshire the number of schools was only 33, one for every 4,900 of the population. In Berkshire it was still lower, only 25, one school for every 6,200 of the population. In Buckinghamshire, it was estimated that there were 36, one for every 4,500 of the population.[1] As already noted, in assessing the cost of farming operations in the nineteenth century no reliable figures can be derived from a mere dependence on the current rates, whether normal or at harvest time, given for the wages and earnings of the able-bodied labourer, for the considerably lower rate paid to juvenile labour and those below the age of 24 must be taken into account.

This juvenile labour can be allocated roughly within four categories: as casual workers employed individually or in small groups (home-based); as members of gangs under a gang master (home-based); as annually hired labourers (home-based); and as annually hired labour away from home (usually provided with lodgings at the farmstead).

In counties such as Oxfordshire, Berkshire and Buckinghamshire with a still-growing rural population, nothing like the gang system existed or indeed was necessary. That none of the special circumstances of the Fen district, where this system mainly flourished and eventually received notoriety and obloquy, were present is clear from the evidence given to the Royal Commission on the Gang System by Mr Nisbet of Thorney, Huntingdon, agent to the Duke of Bedford;

> The only way in which children's labour can be made available... is by working them... in... 'gangs'. The nature of the soil also makes the work of gangs in the fen districts more necessary than elsewhere, owing to the greater number and quicker growth of weeds... Nearly all the farmers in this parish employ the public gangs. One does not, but he has to employ Irishmen instead. Another gets the work done by women only... I have had a wide acquaintance with farming as carried on in different parts of the country, and never heard of any gangs except in these eastern and one or two adjoining counties.[2]

The other three systems were, however, all extensively in evidence in the more populous rural counties. The casually employed went crow scaring in late autumn in the freshly planted wheat fields, and in early Spring in the freshly planted 'Lent' corn, or they might help in bean planting by filling the holes made by the bean 'dibber'. They frequently went weeding these same fields in May and June and in between times picked stones from those meadows earmarked for hay crops.[3] At hay time and harvest time they helped their parents by enhancing their piece work earnings with their 'unpaid' help, making bands and stooking the sheaves. After harvest there was the opportunity to earn a few pence from gathering acorns. Those hired annually who remained at home were usually sons of the annually hired labourers.

At Wick Farm, Radley, on 16 October 1851, Charles Fox was engaged as a shepherd at 10s per week for the first half-year and 11s per week for the last half-year - to shear the sheep at 2s 6d per score and to live in the cottage at Old Farm, paying 1s per week rent. He was given 1s 'earnest' sealing the agreement. His boy also received 1s 'earnest' to seal his agreement, having been engaged at 3s per week with £1 to follow at Michaelmas.[4] The shepherd's boy or cowman's son might not always work in the harvest fields but helped by taking over duties which otherwise might prevent his father taking harvest piece work. An annually hired labourer was working in 1867 on a farm at Begbrooke, Oxfordshire: 'I get 11s a week all the year round; I'm a cattle man and work on Sundays as well. All men here get 11s. I have no Michaelmas money. In harvest I have to go milking in the morning and at night for which I get 6s a week and I go reaping with my boy between times... [My] eldest boy, 13, has 3s a week regular; he drives a team and does any odd job...'[5] About Steeple Barton, stated a labourer, the practice during this period was for 'farmers to keep one carter and all the rest bits of halfling boys at 2s 6d to 4s a week each; the shake of the plough nearly knocks them over'.[6]

Agreements relating to such lads are to be found in the Labour

Book for Manor Farm, Herton, not far from Slough in Buckinghamshire; they contain general details of pay and duties operating over a period from 1846 to 1857.[7] In October 1848, Ned Warden agreed to work as a cow and plough boy, and to make himself 'useful' for 4s per week and lodgings with 30s at Old Michaelmas 1849. John Webb came as a plough boy in 1850 and agreed to work for slightly less, for he was to have 3s 6d per week by £2 Michaelmas Money. By 17 May 1851 he had already drawn 15s of this and in due course received the remainder owing, 25s. The day after Old Michaelmas 1851 Jasper Irving was engaged as second team boy at slightly less wages again - 3s 6d per week and 30s at Michaelmas. In addition, however, to his regular farm duties he was expected to clean knives and shoes on alternate Sundays. Failing to manage not to draw on his Michalmas wage he ended his year's service with 10s the remainder owing, which was duly paid to him on 16 October 1852. William House came to work that October and for his position as 'best' team boy he received the same weekly wage as Irving but 10s extra Michaelmas money. His first purchase was a whip which cost him 5s and having already drawn during the course of the year another 23s, the remaining 12s received at the end of his year's service made him but 2s better off than Iriving who was working under him. Theoretically, House, who was re-engaged at 4s instead of 3s 6d per week but with the same Michalmas money of £2, should have ended his next year's service better off. Unfortunately though he did not have to buy another whip; a pair of trousers bought in August cost him 6s 6d and having already drawn 25s previously he was to find only 8s 6d going into his pocket when paid the remainder of his Michaelmas money on 11 October 1853. On the other hand, Irving the plough boy, also re-engaged, improved his position slightly from the previous year. Though with a slightly lower wage than House, but now raised to 4s instead of 3s 6d per week, and 35s Michaelmas money instead of 30s, he ended the year with 13s, 3s more than the previous year and 4s 6d better off than House. If he had not had to pay 8s for shoes, he might indeed have saved the princely sum of just over £1. That October a lad named Thomas Lovejoy replaced House as best team boy but fared no better than his predecessor money-wise; for with 5s less Michaelmas money (35s instead of £2) he ended his year's hiring on 11 October 1854 with 8s 6d left out of wages due - the precise sum House had saved the previous year.

Meanwhile one Charles Wyse (Wise) had replaced Irving as second team boy on the same hiring terms, but stayed only the one year. He went to Reading on 20 and 23 September 1845, drawing 7s 6d for this purpose, attended Reading Fair, and it must be assumed found there another 'hiring'. His funds when he left Home Farm amounted to 6s, the remainder of his Michaelmas money. Among his purchases during the year which had thus depleted his wages had been a whip costing 3s, 1s drawn on Christmas Eve for he had overspent at the local shop, 3s 6d for a jacket and 2s spent at the 'Revel' on 17 August, presumably the local Church feast.

In turn, William Druce replaced Thomas Lovejoy in October 1854 as best team plough boy also on the same terms, which included cleaning knives and going to church on alternate Sundays. He too was not to stay beyond the year's hiring, despite or perhaps because of the fact that he had not a penny piece of his Michaelmas money remaining at the end of his service. It is revealing to note that apart from the purchase of the essential whip at 4s 6d, all his extra money was spent on clothing - 10s on shoes on 20 January, 2s 6d on shirts and 4s on a flannel jacket in June, 6s for trousers in August, and, finally, a second-hand jacket for 4s and a pair of shoes for 2s 6d bought from the farmer himself.

At Manor Farm the annual change in lads at Old Michaelmas continued, suggesting unsatisfactory working conditions. That other hired men only stayed a few weeks or sought a fresh hiring after their year's service in this period supports this view. Forty years later, in 1891, the conditions at Manor Farm were revealed in a court case held at Slough to be still as bad. Two young men were summonsed for breach of contract under the Employers' and Workmen's Act; one, William Burchell, had signed the following 'remarkable' document:

> I, William Burchell, agree to hire myself to Alfred William and Joseph Reffel for one year as carter at 7s per week for the first half, 8s per week for the second half-year, and £3 at Michaelmas, October 11, 1890, to make myself generally useful at all kinds of work, and to do anything I am asked at any time. In case of any illness or accident I agree to support myself; to be in the stable at 4 o'clock every morning in order to get my horses ready for work at 6 o'clock; to rack up my horses every night at 8 o'clock; to find my own whip, masters to keep it in repair; to get up in the morning when called by the carter; to be in every night by 9 o'clock, except when required to be later by my masters; to clean boots and shoes on Sunday morning.

The report of the case continues:

> Burchell 'left' on December 4, as was alleged without notice, on the ground that he was not supplied with vegetables to eat, which Mr. Reffell had promised him at the time of hiring. He had received, he said, not more than six potatoes during the whole six weeks he was on the farm.
>
> Out of his seven shillings a week he had to find himself food and everything else. If he wanted to warm himself at a fire he had to go to a public house, as no fire was provided. The boiler in which he had to cook was a very rusty one, 'not fit to cook food for pigs in'. Although he had been promised good lodgings, all the lodging he had was a straw bed in a loft up a ladder. Five slept in that room - three in the bed where the witness was and two in another bed. The Bench left little doubt as to their opinion about this

'extraordinary contract'. They said the agreement was a very hard one, and entirely one-sided, for nothing was said about what the employer was to do - nothing about food or lodging. As the law had, strictly speaking, been broken they decided to fine the defendent 6d, but at the same time they rescinded the contract and remitted the costs.[8]

Seeking work away from home was in a sense a natural development; with extending arable cultivation and with an increasing number of horses to be worked and cared for, there was plenty of work available.[9] It was reported in the early 1840s that on most farms in Oxfordshire and on the Berkshire side of the Thames near Oxford two out of three ploughs, and two out of three waggons and horses, were managed by young men under 20 years of age, whose wages varied from 3s to 5s per week, never exceeding and seldom reaching 6s, but sometimes for boys who were hired by the year and who were at work 16 hours a day as low as 2s per week.[10] Somerville had the following conversation with one of these lads on a large farm near Abingdon:

'You hold the plough, you say; how old are you?'
'I bees sixteen a'most.'
'What wages have you?'
'Three shillin' a-week.'
'Three shillings! Have you nothing else? Don't you get victuals, or part of them, from your master?'
'No, I buys them all.'
'All out of three shillings?'
'Ees, and buys my clothes out of that.'
'And what do you buy to eat?'
'Buy to eat! Why, I buys bread and lard.'
'Do you eat bread and lard always? What have you for breakfast?'
'What have I for breakfast? Why, bread and lard.'
'And what for dinner?'
'Bread and lard.'
'What for supper, the same?'
'Ees, the same for supper - bread and lard.'
'It seems to be always bread and lard, have you no boiled bacon and vegetables?'
'No, there be no place to boil 'em; no time to boil 'em; none to boil.'
'Have you never a hot dinner nor supper; don't you get potatoes?'
'Ees, potatoes, an we pay for 'em. Master lets us boil 'em once a-week an we like.'
'And what do you eat to them; bacon?'
'No.'
'What then?'
'Lard; never has nothing but lard.'

'Can't you boil potatoes or cook your victuals any day you choose?'
'No; we never has fire.'
'Have you no fire to warm you in cold weather?'
'No; we never has fire.'
'Where do you go in the winter evenings?'
'To bed, when it be time; and it ben't time, we goes to some of the housen as be round about.'
'To the firesides of some of the cottagers, I suppose?'
'Ees, an we can get.'
'What if you cannot get; do you go into the farm-house?'
'No, mustn't; never goes nowhere but to bed an it be very cold.'
'Where is your bed?'
'In the tollit [stable loft].'
'How many of you sleep there?'
'All on us as be hired.'
'How many are hired?'
'Four last year, five this.'
'Does any one make your beds for you?'
'No, we make 'em ourselves.'
'Who washes your sheets?'
'Who washes 'em?'
'Yes; they are washed, I suppose?'
'No, they ben't.'
'What! never washed: Do you mean to say you don't have your sheets washed?'
'No, never since I comed.'
'When did you come?'
'Last Michaelmas.'
'Were your bedclothes clean then?'
'I dare say they was.'
'And don't you know how long they are to serve until they are changed again?'
'To Michaelmas, I hear tell.'
'So one change of bedclothes serves a year! Don't you find your bed disagreeable?'
'Do I! I bees too sleepy. I never knows nought of it...'[11]

Wages were poor and conditions were poor but it was work. Prior to 1870 it was said 'In agricultural districts children's attendance [at school] was subordinated to the demands of rural labour.'[12] A current labourer's saying in Oxfordshire was, 'I aulus thinks writ'n books, an "praichin", an all sich things as they, be myent for folk as can't work.'[13] An opinion expressed in 1850 regarding the farmer's attitude to education was that '[T]here is no occupation, however slight which does not stand in their estimation, before the school, and they look upon further education, after they are able to go to work, as an unjust deprivation of their labour...'[14] Was it much different after the first Elementary Education Act? Did the implicit threat of compulsory education to the employment of children materialise? If in fact it removed the groups of small figures of children and the lonely 'bird tenters' from the fields in late autumn, the period of greatest labour need at hay and harvest time was very little if at all affected until at least the turn of the century.

Not until 1873, indeed, was it illegal to employ children under the age of eight (36 & 37 Vic. 67 (10)).[15] Not until 1876 was it illegal to employ children under the age of ten, while a specific legal loophole for not sending children to school in remoter rural areas existed in the reasonable excuse clauses of the act (39 & 40 Vic. C79 pt. 1):

> (1) That there is not within two miles... from the residence of such child any public elementary school open which the child can attend; or
> (2) That the absence of the child from school has been caused by sickness or any unavoidable cause.

At periods of greatest need, especially at harvest time, farmers' labour requirements were directly considered by holiday arrangements. Holidays were taken to coincide with the hay harvest in pasture areas; where potatoes were extensively cultivated, a school break in late autumn was arranged, while the great concern for the corn harvest was made explicit by the exact coincidence of the summer holiday period with the commencement of harvest operations, so that this date varied not only from district to district but from year to year in the same district. On a farm near Beccles in Suffolk the start of the harvest between 1813 and 1841 occurred on 19 different days, the earliest commencing 24 July 1822, the latest 28 August 1816.[16] An early and exceptionally short harvest on a 2,000 acre Essex farm in 1828 was completed by 28 July except for the beans.[17] Usually of a month's duration, the harvest was extended about Bradfield and West Berkshire in the 1870s, 1880s and 1890s to five weeks, sometimes to six weeks.[18] This variation in dates created an anomalous position at Lee Common, Bucks, in 1878, and caused concern to the schoolmaster because of its effect on school attendances. His school served the needs of two parish committees, the School Attendance Committee of Amersham Union and that of Wycombe Union. The Amersham Committee had allocated two weeks of the statutory six weeks holiday period to hay harvest and four to the corn harvest; the Wycombe Committee had allocated the whole six weeks to corn harvest. As Lee Common School was closed during the two weeks at hay time, the Wycombe Union children were away from school for those two weeks and quite legally away for the two weeks after the school had reopened on 2 September, thus having eight weeks away from school.[19]

It was not the fault of the attendance officer that nothing could be done to ensure attendance at Lee Common School in this instance. But it is questionable how effective these officers appointed in compliance with the Education Act of 1876 were. A writer to the

'Bucks Herald' signing himself 'Facta non Verbe' was concerned about the results obtained in the previous 18 months after such an appointment by the Amersham School Attendance Committee. 'I feel compelled to answer not only nothing, but absolutely below zero', he wrote in September 1878, 'since people have been both irritated by the interference of the committee, and convinced of the practical uselessness of the officer to enforce attendance at school.'[20] This perhaps was premature judgement but subsequent evidence suggests that his opinion was not so far from the truth. There is ample evidence, indeed, in the school log books of the period to show that 30 years was not long enough to change the habits and thoughts of rural England. It seems an undeniable fact that schooling was acceptable only when the weather was unkind yet it was then that snow or flooding made the passage across the fields or down the lanes to school always hazardous if not impossible. When it was fine the exigencies of work took precedence. References to the weather, to illness, to harvesting and farming pursuits make up a very considerable part of the entries in these official school diaries. However, it was in June, July, August and September - the period of the hay and corn harvest - that school attendance was most affected. The summer break in August was called the 'Harvest' holiday and often entered in large copper plate writing. The dates varied from county to county depending on the commencement of the corn harvest. In Berkshire it became fixed about the third week in July to the first week in September; in Buckinghamshire it was a little later, in Staffordshire as late as the second or third week in August.[21] But whenever harvest commenced earlier or extended beyond the official school holidays, attendance was low. At Bradfield National School, Berkshire, the school broke up for the harvest holiday on 1 August 1873; the entry for 21, 22 and 23 July reads: 'Attendance on these days was limited on account of Harvest.' The entry for the days 28 July to 31 July, just before the school broke up: 'Attendance still very limited.' A month later the school reassembled on 1 September - the entry on that day reads: 'School opened after the holidays... limited attendance.' On the four succeeding days the register remained unmarked because 'attendance very limited'. The following year, 1874, it was decided to extend the holiday period to six weeks to commence on 24 July. This solved the attendance problem after the harvest when it reassembled on 7 September, but the entry for 17 July, a week before harvest commenced, reads: 'decrease in attendance is owing to the commencement of harvest'. However a five-weeks harvest holiday was thought adequate the next year, 1875, from 30 July to 6 September. Mistakenly it seems, for on that day the following entry had to be recorded: 'School should have been reopened today [6 September] but there were so few in attendance that it was closed for another week.'

It might be supposed that in the year that saw the passing of an act setting up school attendance boards the situation would improve. This seems to have been the case for the harvest holiday period remained at five weeks without any adverse comments on attendance either before 28 July when the school closed or after 4 September when it reopened. However, the next year, 1877, despite the precaution of extending the harvest holiday to six weeks from 27 July to 10 September, the entry on that date was: 'reopened after... 6 weeks. The attendance was small.' Hereafter, indeed, the entries are repetitive; with the exception of the harvests of 1884 and 1886 (six weeks) the holiday period became fixed at five weeks, and attendance remained generally low in the week before and after the holiday period, with exceptions in 1879, harvest operations not commencing until 29 August; and in 1880, 1882 and 1883 when no adverse comments were recorded.[22] The entries for other years speak for themselves:

1878 July 22 and following days
Attendance smaller than usual owing to the commencement of harvest operations.
Sept 2 School reopened (over half the school away).
1881 Sept 5 School reassembled. Small attendance.
1891 July 24 Smaller attendance this week; harvest work has now commenced.
1892 Sept 5... several children are still away in the harvest field.

An idea of the relative numbers away can be gathered from juxtaposing an entry of 10 July 1891; '170 children present on Wednesday morning being the highest attendance yet recorded', against the following entries for attendance immediately after the harvest holiday:

in 1887 Sept 5 ... school reassembled present 113
in 1888 Sept 17 .. reassembled ... present 96
in 1889 Sept 2 ... reopened ... present 74
Sept 10 .. small attendances throughout the week. Harvest not yet complete.
in 1890 Sept 8 ... reopened ... only 80 present
Sept 12 .. a number of children not yet returned.

There is no evidence that the attendance officer was particularly concerned at the situation existing at Bradfield National School. There was certainly nothing like the frustration on the part of the head teacher that was shown over a similar situation at Wall Street Village School near Lichfield, Staffs, where the attendance officer appeared one day in 1881 (6 June) and did not reappear until three years later (28 April 1884). That the school inspector was concerned though he proposed no remedy is shown by his report dated 25 November for 1885: 'there should be more evidence of precise and energetic school keeping than is now traceable'. The following entries relating to attendance at hay time as well as those later in the summer already enumerated must have influenced his remarks:

1879 June 6 Small attendance throughout the week (Whit Monday and Tuesday had been holidays)
1884 June 26 The Haymaking having commenced several children were absent this week.

A previous entry on 20 June that year indicates a somewhat dispirited head teacher for he complains that '[T]he general attendance is most irregular and it seems no use sending names to A. Officers as they do not come any better'.[23]

Absenteeism was a widespread problem and the comparative absence of prosecutions for absenteeism was certainly not due to satisfactory attendance. Children stayed away to gather fruit, to glean, to gather acorns;[24] when the opportunity arose they took more permanent jobs in the summer months. So long as parents needed the economic support of their children's earning power, and farmers needed this extra labour, the influence of teachers in maintaining attendance was negligible, the attendance officers' work ineffective, the school attendance board's attitude lenient. When Wall Street Village School near Lichfield, Staffs, reopened on 17 September 1888 after the harvest vacation, '[S]ome of the children were yet in the [H]arvest field. The Standard V boys were attending "very badly". The harvest [was] not nearly over and they [were] wanted to "lead" the horses.'[25] Farmers saw schooling as a drain on their immediate labour supply and educated children as a threat to their future labour supply. Parents faced with school fees and a loss of their children's earning power, however small, were inclined to see learning as irrelevant to their child's future in a rural occupation. Learning about the practical details of farming was sensible training for a future occupation. An Addington mother's request in May 1885 that her boys should have a morning off school to see sheep washed was an understandable one. That this request like so many similar ones was not refused is also understandable.[26] Teachers had rarely either the power or the social standing to insist that education was something more than a training for work. As far as parents and most farmers were concerned the dilemma of rural education was an academic one; work took precedence for reasons of either economic necessity or economic gain.

The log book of Addington School, a Buckinghamshire village school near Winslow, shows the same pattern of poor attendance. The comments and reasons given are not so complete, but any school log book of the period would provide the same general picture if consulted and a veritable calendar of country activities could be reconstructed from their close study. But neither at Bradfield National, nor at the Village Board School, nor at Addington were there any records of prosectuions for non-attendance (at least not until 1893 at Bradfield National, 1900 at Bradfield Village and 1896 at Addington).[27] At East Ilsley, Berkshire, the school board met on 5 August 1895 and irregular attendance was considered. At the following meeting on Monday 23 September, 'it was not considered necessary to take out any summonses for irregular attendances'.[28]

There were of course occasional bursts of action. Five Binfield men and two Warfield men were fined 5s each, and two Winkfield men were fined 5s and 3s 6d respectively by Berkshire magistrates in August 1882.[29] In a climate of apathy and against the fact that there existed in 1886 only 718 school attendance committees and 1,908 school boards in England and Wales, with about two-thirds of rural England still under the exclusive supremacy of the denominational system in primary schools,[30] it is not surprising that the extent of prosecutions was minimal in many districts. An opinion expressed in 1895 gave the following reason for this situation:

> By the more prosperous villagers he [the village schoolmaster] is frequently regarded as an expensive and unnecessary luxury, a burden on the rates and tax payers. The more conscientiously his duty is performed as regards the regular attendance of his scholars, the less is his presence and office relished, alike by employers and the parents of his charges.[31]

Of the three counties of Buckinghamshire, Berkshire and Oxfordshire, Buckinghamshire seemed the least unconcerned. Fines of up to 5s were imposed on six parents prosecuted by the Eton School Attendance Committee for not sending their children to school regularly in 1886.[32] In July 1889 the attendance officer reported to the Winslow Attendance Committee that 'at Stewkley the attendance was below average owing to many of the children being temporarily employed, while the younger ones were kept at home to carry rations to the hay field'.[33] In August he reported that fines imposed on Whitchurch and Quainton parents had remained unpaid; also at Stewkley where more summonses had been taken out against parents than in any other part of the union.[34] Seven Stewkley parents had to appear before the Linslade Magistrates on 12 August where they were charged with neglecting to send their children to school:

> In every case the attendance had been very bad, some of the children having been almost totally absent during a term of three months, and the excuses of three defendants only who appeared were very trivial. The defendants were fined 5s in each case. The Chairman said parents were foolish who did not comply with the law passed in the interests of their children. A child sent to school regularly, unless a fool, would pass the 4th Standard and be freed from further attendance at the age of ten.

At the same petty sessions:

> Alfred Baker, Thomas Smith, Berry Staples, and Emma Brown, all of Wing, were summoned by Mr. W. F. Broom, on behalf of the School Attendance Committee of the Leighton Buzzard Board of Guardians, for neglecting to send their children to school; and Dan Rickard of Wingrave,

> was charged by Mr. J. N. Hodgkinson, on behalf of the School Attendance Committee of the Aylesbury Board of Guardians, with the same offence. The wives of the three first-named appeared; Brown did not appear, nor was he represented. Mrs. Staples pleaded that her son played truant; she had always sent him to school, with his money for fees; she had never heard any complaint, nor had she ever seen the Attendance Officer. Mr. Broom stated, in reply to this, that he had twice recently seen Mrs. Staples; the usual notices had been served, and defendant had been summoned before. Mrs. Smith said that if she sent her boy to school one day he would 'come home bad the next'. In Brown's case it was stated that Mrs. Brown had positively refused to send her child to school, and told the Attendance Officer that he might take out summonses as often as he liked. The other defendants did not appear. Fines were imposed of 5s in each case, the Chairman remarking that the Bench had no power to make the penalty higher, or they would do so, since each of these cases cost the county 7s 6d., and he did not see why the county should pay for people neglecting to send their children to school.[35]

The attitude of Emma Brown was not altogether exceptional. William Grover of Hughenden had in 1878 kept his boy away from school to assist him in his trade of hay binding; when summoned he 'persisted in his determination to keep the boy at work'.[36] Despite the fact that at Stoke Goldington, four miles north-east of Newport Pagnell, attendances were reported as 'very bad, worse than in any other part of the district', and at Newport Pagnell nine parents were summoned over school non-attendance on 12 August 1889, the Winslow Board of Guardians was told by the attendance officer at a meeting on 14 August that attendance at the schools in that area was very good. Stewkley parents, indeed, were presented as exceptional in their behaviour, though fines were generally paid upon legal proceedings being taken: '[On] one occasion half a dozen parents went to gaol.'[37]

It seems quite clear that farmers in general by continuing to employ school children were aiding and abetting, even discouraging parents from sending their children to school. A 'good deal of pressure is put on parents by the farmer to obtain the labour of children for the sake of saving the wages of a man' was the belief of a Middle Claydon villager before 1870.[38] As august a village personage as the erstwhile chairman of Coombe Parish Council, Oxfordshire, was prosecuted as late as September 1898 'for illegally employing a boy... contrary to the Elementary Education Act' and was fined 10s at South Wootton Petty Sessions. A most astounding case had occurred nine months earlier at East Ilsley in Berkshire. A parent had been successfully prosecuted for not sending his boy to school since 9 July 1897. The fine of 2s 6d had been promptly paid in November but the boy continued his absence. At a meeting of the school board on 27 December 1897 it was decided to take out a further summons against the father. The dilemma that then faced them was explained in this extraordinary letter composed and agreed to by the board at a meeting on 6 December 1897, duly drafted by the clerk and sent to the Lords of the Education Department:

> This proceeding [a further summons against the father] however, though necessary, can hardly be effective. For it is quite an open secret in the Parish that the boy is being regularly employed by Mr. Francis Stevens, the Manor Farm, East Ilsley, who is the largest farmer in the village. All the Members of the Board present at the Meeting expressed themselves as being very averse from taking legal proceedings against Mr. Stevens, as this would cause very serious ill will, and much personal feeling, in this small village, in fact that course is practically impossible... Would it not be possible for the Education Department, as an impersonal and independent body, to write to Mr. Stevens, and to point out to him that by employing a boy who ought to be attending school, he is defeating the object of the Education Acts, and rendering himself liable to punishment. Without some such help from outside, the Board feel that they are not equal to the requirements of this case...[39]

At a special meeting on the 27 December the reply dated 17 December 1897 from the Education Department was laid before the board for consideration:

> Rev. Sir, Adverting to your letter dated the 6th instant, I am directed to state that my Lords cannot consider that the circumstances explained by you are such as to warrant your Board in taking no action in this matter. If the School Board neglect to take proceedings against the employer under the Elementary Education Act, 1876, on the grounds stated in your letter, it will probably be necessary for Their Lordships to declare your Board in default and to appoint other persons to discharge their duties (see section 27 of the Act). You are, no doubt, aware that the Education Department may direct that the other persons so appointed may be paid for their services (see section 65 of the Elementary Education Act, 1870). My Lords cannot undertake to communicate with Mr. Stevens in regard to this matter...

These two letters were then discussed, but a decision was postponed until a third meeting on Monday 3 January 1898. At this meeting, two further letters were read; both headed 'The Rectory, East Ilsley, Berks' dated '3 January 1898'. One contained the resignation of the chairman of the board signed by the Rev. Thomas R. Terry, MA, FRAS, Rector of East Ilsley and late Fellow and Tutor of Magdalen College, Oxford, who had held this office for the previous 13 years; the other read: 'Gentlemen, I hereby beg to resign the post of Honorary Clerk of the East Ilsley School Board. I have the honour to be Gentlemen, Yours very truly, Thomas R. Terry.'[40] At a board meeting a month later, it was reported to the newly elected chairman that proceedings had been taken against Mr. Stevens and he had been duly convicted and paid the costs.

H. C. Darby suggests that one of the contributory causes of the agricultural depression was, in the Fens, the difficulty of obtaining juvenile labour.[41] W. H. C. Armitage suggests that radical leaders suspected '[T]he controllers of... country schools [the squire and the clergyman]... of being lukewarm on the education of future farm labourers,' and he quotes Sir John Gort repeating a contemporary opinion:

> the farmer and the squire are no friends to elementary education. They associate agricultural depression and high rents with compulsory education and they grieve to pay for that teaching which deprives them of servants and families their labourers with wings to fly from the parish. On the other hand, the labourer has not yet learned the value of education. The earnings of his children are important to him and the present shilling obscures the future pound.[42]

It would seem that at least in Oxfordshire, Buckinghamshire and Berkshire all the protagonists in one way or another ensured that whenever juvenile labour was needed it was usually forthcoming.

Table 4.1: Offences Against the Education Acts, England and Wales, 1870-1901[1]

Year	Total	Berks	Bucks	Oxon
1873	6,693	- (117)	(-)	11 (10)
1874	15,036	- (110)	39 (-)	33 (28)
1875	21,386	- (148)	32 (-)	38 (27)
1876	25,129	100 (92)	25 (-)	42 (33)
1877	23,356	110 (89)	13 (4)	34 (25)
1878	40,836	277 (129)	63 (2)	183 (25)
1879	49,845	370 (109)	226 (2)	320 (42)
1880	55,696	475 (166)	330 (16)	366 (88)
1881	67,352	798 (284)	479 (39)	319 (34)
1882	83,474	887 (276)	718 (43)	521[2] (74)
1883	94,274	904[2] (282)	819[2] (45)	494 (102)
1884	86,027	724 (283)	685 (84)	405 (150)
1885	76,173	629 (271)	523 (74)	330 (100)
1886	67,093	502 (264)	350 (39)	220 (111)
1887	76,265	614 (326)	425 (102)[2]	305 (107)
1888	76,589	592 (359)	347 (101)	272 (108)
1889	80,519	666 (331)	319 (41)	381 (157)
1890	87,439	650 (300)	287 (27)	354 (138)
1891	96,601[2]	824 (424)	470 (43)	425 (189)
1892	86,149	853 (412)	390 (0)	309 (122)
1893	63,015	674 (291)	260 (6)	200 (87)
1894	62,494	639	310	181
1895	59,737	525	371	213
1896	67,859	583 (317)	326 (10)	189 (56)
1897	71,518	799 (392)	479 (28)	182 (71)
1898	79,464	771	521	246
1899	89,432	768	559	226
1900	89,567	737	469	198
1901	78,514	676 (394)	331 (47)	203 (58)

Notes: [1] Urban offences in parentheses when differentiated, but included in county totals.

[2] Highest annual totals underlined.

Table 4.2: Education Act Offences, Proportion of Persons Tried Per 100,000 Inhabitants, England and Wales, 1874-94

Year	No.
1874-8	103
1879-83	272
1884-8	278
1889-93	285
1893	212
1894	208

Source: A & P 1896 XCIV, Judicial Statistics for 1894, p. 16.

Table 4.3: Education Act Offences, Annual Average of Prosecutions, England and Wales, 1872-96

Year	No.
1872-6	17,061
1877-81	47,417
1882-6	82,008
1887-91	83,483
1892-6	67,851

Source: A & P 1898 CIV, Judicial Statistics, pp. 36-7.

NOTES

1 'Morning Chronicle Supplement', 28 Dec. 1849.
2 'Ag. Gaz.', 20 Apr. 1867, p. 418.
3 PP 1867-8 (4068-I) XVII, RC Employment ... in Agriculture (1867), App. pt. II, p. 518.
4 Reading Farm Records, BER 13/5/3, Labour Book, Wick Farm, Radley, Berks.
5 PP 1868-9 (4202-II) XIII, RC Employment... BJ, p. 570 (5b).
6 Ibid., p. 570 (3f).
7 Reading Farm Records, BUC 1/5/1, Labour Book, Manor Farm, Horton, Bucks.
8 'The Land Worker', vol. 28, no. 337 (June 1947), p. 11 (extract from the 'Iron Workers' Journal' of 1891). Note: available work was given to young lads rather than young men; to men with large families rather than to men without, thus saving on wages and on the poor rate; cf. Supplement 'Morning Chronicle', 24 Dec. 1849, 'it's made a matter of pounds, shillings and pence... not of rewarding them that faithfully serve them' (Thame labourer).
9 See 'Oxford Chronicle', 17 Sept. 1853; ibid., 1 Oct. 1853. Nine boys under the age of 18 had all started work between the ages of 7½ and 9½ in the vicinity of Chipping Norton.

10 A. Somerville, 'The Whistler at the Plough' (Manchester, 1852), p. 142.
11 Ibid., pp. 142-3; but for 'satisfactory' conditions at Bowden Park, near Chippenham, Wilts, see Thomas Dyke Ackland, On Lodging and Boarding Labourers as Practised on the Farm of Mr. Sothern MP, JRASE, vol. X (1849), pp. 379-81. Wages £4 to £8 10s, purchase of own clothing, but food provided (5s 1½d per week).
12 W. P. McCann, Elementary Education in England and Wales on the Eve of the 1870 Education Act, 'Journal of Educational Administration and History', vol. 2, no. 1, (University of Leeds, 1969), p. 26. The common tasks were bird scaring, stone gathering, straw-plaiting and acorn picking (PP, 1871 XXII, pp. 210-11).
13 Angelina Parker, 'Supplement to Oxfordshire Words', p. 32 (113).
14 'Ag. Gaz.', 27 Apr. 1850, p. 267; cf. R. Trow-Smith, 'Society and the Land', (1953), p. 140. Not until the Education Act of 1876 were the infants of the countryside taken from the cold wet fields into the village classroom.
15 An act to regulate the Employment of Children in Agriculture.
16 N & Q, 2nd Ser., vol. 81, 18 July 1857, p. 57.
17 Ibid., 2nd Ser., vol. 79, 4 July 1857, p. 8.
18 Berks. RO, C3EL9/1, evidence from Bradfield National School Log Book.
19 'Bucks Herald', 7 Sept. 1878.
20 Ibid., 28 Sept. 1878.
21 See Reading University Library Archives, Wall Street School Log Book, near Lichfield (uncatalogued).
22 Bradfield National School Log Book. Note: The situation in 1885 cannot be clarified since the school was not reopened until 3 November because of an outbreak of measles.
23 Wall St Log Book (uncatalogued).
24 Bucks RO, E/LB1, Addington School Log Book. E.g. '1886 July 23 ... several children have been irregular in their attendance... Fruit picking is the principal reason given; 1888 Sept 21... The Cullums and Whites absent... gleaning'; Bradfield National School, '1890 Nov 7... A large number of children away getting acorns.'
25 Wall St. School Log Book (uncatalogued).
26 Addington Log Book, 26 May 1885.
27 See Table 4.3, p. 73, for number of prosecutions under the Education Acts.
28 Berks RO, C/EB5, Minute Book, East Ilsley School Board, ff.162.
29 'Reading Mercury', 12 Aug. 1882; ibid., 19 July 1893. Two Hartley Row men and a Mattingley man prosecuted.
30 'Windsor & Eton Express', 12 June 1886, extract from School Boards and Attendance Committees - yearly list.
31 Joseph J. Davies, The New Minister of Education and his Work (Sir John Gort), 'Westminster Review', vol. CXLIV (1895), p. 335.

32 'Windsor & Eton Express', 7 Aug. 1886.
33 'Bucks Herald', 13 July 1889.
34 Ibid., 3 Aug. 1889.
35 Ibid., 17 Aug. 1889.
36 Ibid., 23 Nov. 1878, High Wycombe Petty Sessions, 15 Nov.
37 Ibid., 17 Aug. 1889.
38 See PP 1867-8 (4068-I) XVII, p. 552.
39 Berks RO, C/EB5, Minute Book, East Ilsley School Board, ff.180-1.
40 Ibid., ff.182-3.
41 H. C. Darby, 'The Drainage of the Fens' (Cambridge, 1968), p. 246, 'due to the Education Acts'.
42 W. H. G. Armitage, 'Four Hundred Years of English Education' (Cambridge, 1970), p. 182.

5

THE IRISH HARVESTERS

> before the days of field machinery... the great English crops could never have been gathered and saved... had not Ireland come to help.
>
> T. Mackay, 'Reminiscences of Albert Fell' (1908), p. 144.

Broadly speaking, three factors decided the initial destination of Irish migrant harvesters: their point of departure; work opportunity; and wage rates. Assuming that work at any price was the objective, their most obvious destinations would be those corn-growing counties when in a period of ever increasing production there might be developing an absolute as opposed to a relative shortage of labour. Handley points out that wages for migrant workers could vary: 'when supply much exceeded demand... as low as a shilling a day with food and lodging provided... when the harvest was in full swing... [they] might reach two shillings or two and sixpence. At the end wages fell again...'[1]

In the south there was an indigenous short-distance migratory flow already in existence. For example, North Hampshire labourers about Tadley harvested in the Reading area of East Berkshire; villagers in the Vale of Shaftesbury moved out of Dorset into the Fordingbridge area of Hampshire at harvest time; West Berkshire villagers and inhabitants of North-West Buckinghamshire found harvest work in South Warwickshire;[2] Farnham villagers in Surrey had a traditional annual exodus 'down into Sussex' for the corn harvest.[3] In June each year the Middlesex hay fields provided work opportunity and early relief from winter underemployment for some of the rural populations of West Buckinghamshire and North-East Oxfordshire.[4]

The established centres south of Caird's line were Cambridge, Warwick and the hay fields of Middlesex. The Irish harvesters came in numbers for the early hay harvest in Middlesex. About the last week in July they congregated in Cambridge 'in vast numbers' where in the late 1840s there were 13 separate lodging houses for their use.[5] They were in the wheat fields about Warwick in the late 1840s and at weekends brawled with the railway navvies.[6] Those who landed at western ports such as Bristol had a comparatively short distance to travel, but those who landed at Liverpool were encouraged to travel the much longer distance south through Lincolnshire and Cambridge by the certainty of this earlier harvest in Middlesex. As the corn ripened in the environs of London, harvest work became available in Sussex, Kent and Essex for those who preferred not to retrace their steps immediately. Otherwise they commenced their return journeys west along the Bath road. The scene is set by Jefferies:

> outside London the wheat [is] already bronzed... and the reapers at work; thirty more miles west it is only yellow, and reap-hook has touched not it; thirty miles further and it has barely turned colour at all.[7]

Or they returned northwards through Hertfordshire,[8] taking the harvest as they went, for in each county further west and north the harvest was slightly later, giving the opportunity of taking two or more harvests as they travelled nearer to their original points of departure.[9] However, despite the fact that the Irish were present in numbers on the eastern boundaries of Oxfordshire, within striking distance of Buckinghamshire from the north and free to penetrate its southern boundaries after hay making in Middlesex, both these counties were not in the harvest 'flow paths' established over the years. The numbers of migrant harvesters moving south and west were considerably less than the main flows, which according to Collins were in the direction of Surrey, Sussex and Kent (with the added attraction of the hop season), and northwards into Hertfordshire, north-eastward into Essex.[10] The main flows were slowly drying up by the 1860s.[11]

Evidence, however, from the late 1860s suggests that though numbers were declining in the south the Irish were still very active in certain counties; in Northumberland, Yorkshire and Cheshire in the north; further south in Shropshire together with the Welsh reapers; in Worcestershire, Warwickshire and Gloucestershire where one-quarter of the wheat crop was cut with the sickle; in the south-east in Essex where the whole of the harvest was taken 'in companies' with 11 acres the average for each man;[12] in the Fens where the sickle was abandoned for the scythe and where Hall states that the Irish were still cutting by hand into the twentieth century. In the early 1880s they were in Surrey, coming 'year after year to the same barn for the hoeing and the harvest travelling from the distant West'.[13] Richard Jefferies also observed in 1877 that 'they must be counted by thousands', and in the apocalyptic year of 1879 that 'the Irish hay makers were wandering about Wiltshire unable to find work'. 'There is a vast amount of work crying out to be done... thousands of acres of corn upon which no clearing work could be done... thousands of acres of grass waiting to be cut.'[14]

By the 1880s the weight of evidence suggests that the main body of the Irish no longer travelled further south than Warwickshire and North Cambridgeshire in any numbers, and sought and found harvest work mainly in the northern half of England. It is suggested that the certainty of work rather than actual harvest rates determined the flow paths of migrant labour. Kerr, however, suggests that there were few Irish in Dorset because of the low wages.[15] Rather it might have been because the Irish radiating out from Middlesex had no reason to travel further than Berkshire, Wiltshire and Hampshire to find sufficient work to last out the harvest period. Although low annual hiring rates indicated a winter surplus of labour and potentially no absolute shortage of labour at harvest time, uncertainties of weather, state of crop and the immediate availability of labour, imponderables varying from year to year, from farm to farm, were such that in any one district what might seem an absolute shortage one year might be a surplus of labour another, or a relative shortage occurring during the actual harvest operation, either because of inefficient organisation, or inefficient work performance, or upstanding fields of corn being flattened overnight by sudden storms. It is understandable, therefore, why the Irish did not choose to radiate out into Oxfordshire and Buckinghamshire, for both were counties of low earnings and a rising labour force, suggesting something approaching adequate provision of labour for the harvest period. Because of the uncertainties of harvest there could be no optimum labour force; farmers wanted a minimum of workers in winter and at harvest time to engage 'all available labour'.

Individually or in small groups the Irish harvesters could deviate from their familiar routes and find work, especially in the middle years of continuing expansion in the corn-growing counties. Individual Irish were about Eynsham in August 1864.[16] In late October 1857 Irishmen were still working on the Ditchley Estate, potato picking at 10s per week.[17] In the 1860s Heath lists Oxfordshire as one of the counties where Irishmen were prone to find harvest work, while in 1865 a farmer was complaining that his regular Irish harvest helpers had not returned that year,[18] and an advance report of the harvest prospects in the Cholsey district intimated that though the wheat crop was light on some lands, an average crop was expected, and that '[T]he scythe and the sickle will be brought into requisition in a few days'[19] - a sure if inferential indication that there were Irish harvesters about. They were in the Banbury district in the harvest of 1873,[20] and they were in harvest fields about Barford as late as 1889, if the reports can be relied on, for 'everywhere around the sickle [had] been brought into requisition and one field after another of ripened corn [had] been laid low'.[21]

As late as the 1880s Flora Thompson suggests that the Irish harvest labour was available to help out in North-East Oxfordshire. Unlike the term 'Mickies' which was applied to the Irish in East Anglia, where they had a notorious reputation for being 'quarrelsome and insubordinate',[22] they were in Oxfordshire by this time considered rather figures of fun:

> 'Here comes thay jobbering old Irish' the country people would say, and some of the women pretended to be afraid of them. They could not have been serious... [for] all they desired was to earn as much money as possible to send home

> to their wives... for themselves to get drunk on Saturday night, and to be in time for Mass on a Sunday morning.[23]

If this was unnatural docility, there were conflicting opinions as to whether the English or Irish were most to blame for open hostility. On the one hand the Irish in East Anglia seemed not to wish to be provoked for they 'all desired to be locked up at night safe from the English' and would very often ask the same favour when a strange Irish Gang was engaged.[24] Yet the English 'declared that heaven would not be heaven if the Mickies were in it'. On the first Sunday in August 1848 at Warwick:

> a tumult arose in a part of the town called 'Saltisford' where at this season of the year, the Irish labourers usually congregate... not less than three hundred [were] present [in a fight]... against the navvies... women [were] also taking part throwing stones.[25]

A week before, a number of Irish assaulted a shop assistant at 11.30 on the Saturday evening;[26] that Friday a dozen Irish had appeared at Warwick County Court in a state of intoxication; they had hooted out 'There go two bloody Saxons' and assaulted them.[27] One evening in the second week in August the same year three Irish who were refused beer by the landlord of the Lamp Tavern, Warwick, threatened to come back and 'cut [his] head off', and subsequently returned with a sickle and assaulted him.[28]

The Irish in their turn were attacked, mostly by the 'navvies'. That August indeed was a rowdy month: 'scarcely a week passed in Warwick without some conflict with the Irish reapers', the navvies frequently being the aggressors. In August 1848 an Irishman died following 'a severe beating from the navvies'. During that month several Sunday evenings suffered disturbances at Dunchurch near Rugby. At Harborough Magna an Irishman who had harvested for the previous 25 years was attacked by a navvy. Later in October an Irish labourer was attacked by navvies near Clifton.[29] Even so, the Irish were castigated by the Leamington press as a 'set of low Irish... displaying their Milesian turbulence'. 'These caitiffs make themselves intoxicated... and sally forth to bully and quarrel with each other.' The editor of the 'Courier' waxed strong:

> It is really too bad that fellows who take the bread from the mouth of our own surplus population should attempt... outrage[s] at our very doors... the attention of the constables [should be drawn] to the doings of these emeralders who appear to be stragglers from the main hordes that have ravaged England like another eruption of the Huns of Attila... the vagrant labourers and beggars we meet with wandering through the country are about the worst specimens of barbarism we could desire to meet.

The occasion for this outburst had been a brawl earlier in July between a party of English navigators and the 'ruffianly' Irish at Leamington.[30]

The navvies were to depart from Warwickshire but the Irish kept returning. They did not improve their reputation. At harvest time in 1860 during the first Sunday in August 'the peace and quietness [of Stratford on Avon] were much disturbed... by a party of Irish labourers parading the streets in a drunken and riotous manner insulting the inhabitants and threatening the police from an early hour in the morning'.[31] Earlier in July cases of drunkenness and Irish women brawling in Warwick were reported.[32] Later in August two English labourers were involved with 30 or 40 Irishmen fighting in the kitchen of the Nelson Inn, Longbridge, on a Saturday night while at the same time a father and his son with three other Irish, all drunk, 'were creating a disturbance on the Longbridge Road'. On the Sunday morning, a week later, the Red Horse public house was the scene of trouble - a single policeman attempted to stop fighting which had broken out. When immediately surrounded by 33 Irishmen, on blowing his whistle, he was struck at with a sickle which tore his fingers.[33] In 1880 they were still coming and still getting into drunken brawls.[34]

Between the impression of the Irish provided by Flora Thompson and Sir John Sebright's statement in his evidence to the Select Committee on Emigration in 1826 that 'their conduct has been invariably most exemplary'[35] there was a span of 60 years. Jefferies adds credence to their view: '[O]n the whole they are sober, and rarely quarrel except among themselves, or if they have women with them... They are civil and obliging and quite honest.'[36]

The reality of their behaviour, however, exposes not a gap in time but a 'gap' in credibility. Regarding their living conditions, exposure is also perhaps the right word for a drunk and disorderly Irishman on a Sunday night in August 1860 who proceeded from the Seven Stars Inn, Warwick, to his lodgings and there paraded himself nude. This personal exposure led to the revelation in court next morning that eight persons were sleeping in his room 'in a state closely resembling that of pigs'.[37] But it is clear that crowded sleeping conditions were usual, not exceptional. A Stoke Poges farmer in the 1860s related how dealers who brought peas from the local farmers employed the Irish to pick them. '[W]hole families come and sleep anywhere they can sometimes 30 or 40 in a barn the families sleeping together. It is said very little immorality takes place owing to the greater chastity of the Irish girls.'[38] Indeed, between crowded lodgings and barn accommodation there could have been no alternative, and in the 1870s Richard Jefferies formalises Irish 'barn living' as a customary practice of the open countryside.

> One day I observed a farmer's courtyard completely filled with groups of men, women, and children, who had come travelling round to do the harvesting. They had with them

> a small cart or van - not the kind which the show folk use as moveable dwellings, but for the purpose of carrying their pots, pans, and the like. The greater number carry their burdens on their backs, trudging afoot... [I]n a corner of an arable field... the men of the family crowd over a smoky fire... These people are Irish, who come year after year to the same barn for the hoeing and the harvest, travelling from the distant West to gather agricultural wages on the verge of the metropolis... When the corn is cut these bivouac fires go out... next season the smoke will rise again.[39]

If they were variously described as 'Mickies', 'Paddies', natives of the Emerald Isles, 'emeralders', 'Milesians' or 'Goths', their work capacity and ability were also variously assessed.[40] A Radford farmer thought them 'not worth so much as Englishmen' but was reprimanded by a magistrates bench for such derogatory remarks.[41] Redford indeed states that 'they were good reapers but not fit for anything else'.[42] MacKay agreed with him: '[they] were masters of only one tool, the sickle, not the hook, not the scythe, and were positively dangerous when using a fork'.[43] But when two Irishmen in September 1848 at Stratford County Court claimed payment for some reaping and a Loxley farmer complained that the work had been 'badly done', a witness stated that he had 'never seen a crop better cut'.[44] If they were good reapers in the hay fields of Middlesex, they were still strangers to the technique of scything in the earlier years of the century and went there as hay makers.[45] By the 1860s, however, they were corn harvesting with the scythe both in Ireland and especially in the Fens where the corn crops betrayed the excess of nitrogen in the soil, were rank and full of growth with an enormous yield of straw, and whether twisted by storm damage or not called for a great expenditure of hand labour.[46] They were in demand in the Scottish potato fields from the 1870s,[47] while '[t]hey always asserted', states McKay, that 'their last employment before haytime was in Lancashire shaking gwarner [guano] on the pertaters'.[48]

If the migrant flows were to decline from the late 1850s in the south the decrease was not felt immediately or even for a considerable time for numbers of Irish harvesters chose to remain each year in England. Taking any work they could in the winter months they continued to move out into the Middlesex hay fields while contributing largely to the annual exodus into the hop fields of Kent. This reservoir of hidden labour is made explicit in a contemporary analysis of the population of the capital city:

> the practice was for the Irish to come over... to help in harvest... of late they have remained here... great numbers leave London during the hay season, the hop season or the harvest and return when these have ended...[49]

The annual exodus was to continue[50] though there is no way of knowing the numbers of these now indigenous Irish who annually crossed an occupational barrier to join those who annually crossed the Bristol Channel and the Irish Sea into the harvest fields adjacent to London.

Between 1860 and 1880 the full effect of massive Irish emigration to America was felt so that numbers declined to somewhere in the region of less than 40,000.[51] It must be assumed that this emigration was in part responsible for the decline in numbers coming to harvest in the southern counties, although the natives of these counties still flocked to Dublin entering England via Liverpool. With the declining rural populations concentrated in the northern counties of England, the Irish migrants could more readily find work there not only at harvest time but earlier and later in the year and, critically for the south, much nearer their port of entry. Thus, while the direct flow south through ports below Liverpool diminished, the flow south from Liverpool itself was curtailed by being absorbed more immediately in the labour demand in northern and midland counties.

In the process the Irish were to undergo a second transformation. Described as travelling harvesters for more than half a century, many might now begin to be termed semi-permanent farm workers.[52] What, however, is not explained is Jefferies's statement that the Irish were still in considerable numbers about Wiltshire and Surrey in the late 1870s. As he puts it:

> [It] would be interesting to know if the comparative cessation of emigration to America has caused any increase in the numbers who have visited us this year [1877], they must be counted by thousands.[53]

Yet if the flow from the west of England was greater than available figures suggest, it must be inferred that the direct flow from west to east must have dried up considerably by 1880, since by then the main source of migrant labour came almost entirely from the north-west of Ireland with Liverpool the main point of entry. Migrants from the west might be numbered in hundreds after the late 1860s, certainly not in thousands. Between 1880 and 1900 however, Scotland was the most affected by the decline in numbers. Using Jackson's lower figures as pointers and assuming that points of departure indicated Scotland as their destination, 36,514 Irish were harvesting there in 1841.[54] Using Handley's figures from 1880 to 1900, out of a total of an estimated 38,000 arriving in Great Britain in 1880 only 3,771 actually went to Scotland. Of the lowest number arriving in the two following decades, 20,000 in 1889, only 1,231 went to Scotland. By 1899 the numbers had grown steadily again to around 31,000 for the whole of Great Britain and a slightly higher proportion went to Scotland, 4,253, representing approximately 14 per cent of the 1899 influx compared with slightly less than 10 per cent in 1880. From these figures, it is clear that the number of Irish in England was something like 10,000 more in the early 1880s than in 1841, standing at roughly 30,000. If only one-tenth moved further south than the midlands, then Jefferies might have been wrong only in regard to the direction from whence they came, not mistaken as to their numbers, while many may have been the London-born Irish.

Of the final peak figure in migration into Britain, estimated as 32,000 and occurring in 1900, there is no way of knowing their actual distribution. That their numbers in the southern counties of England must have been small is fairly clear from the position in England as a whole summarised by an extract from a report of A. Wilson Fox:

> The men from the West of Ireland go to the Northern and Midland Counties of England,... Very few of the men who go to England go further south than... Warwickshire, Lincolnshire, and North Cambridgeshire. Some also go to... Northamptonshire, Middlesex, and Hertfordshire. A very few are said to be sometimes found in parts of Worcestershire, Bedfordshire, Buckinghamshire, Surrey and Sussex. In former years... they came in considerably larger numbers to some of the counties referred to... reports from Berkshire, Oxfordshire, Huntingdonshire, Kent, and Herefordshire, state that they used to come there, but have now ceased to do so... owing to the introduction of machinery at harvest, and... the smaller acreage of grain crops grown... A good many men manage to get two harvests by going further north when they have completed one in a more southern county.[55]

A final conclusion that might be suggested from the rather confused figures is that although the peak period of Irish migration for harvest work was around 1850 for Great Britain as a whole, the massive influx into Scotland and the fairly quick decline of much smaller numbers in the south of England have masked the fact that the peak period for England when considered separately was quite probably in the late 1870s, concentrated in the north and midlands.

Figure 5.1: Irish Migrant Harvesters to Great Britain, 1880-1900

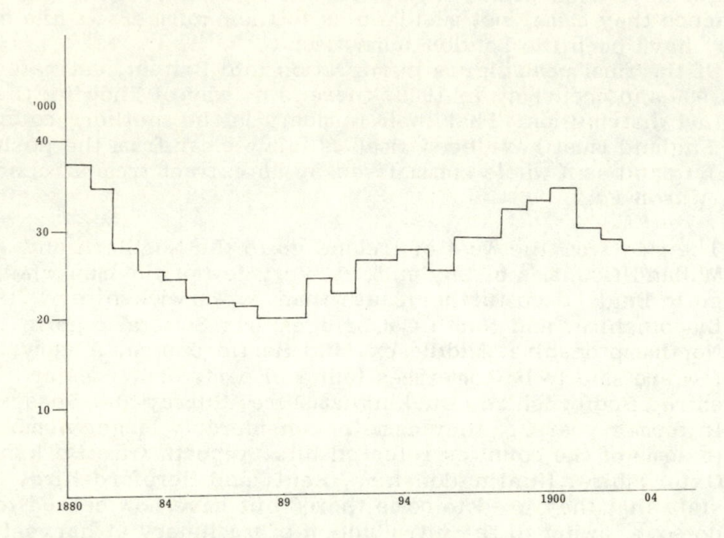

Sources: J.E. Handley, 'The Irish in Modern Scotland' (Cork, 1947), p.172, Table II; John Archer Jackson, 'The Irish in Britain' (1963), p.193, App. Table XI.

NOTES

1 J. E. Handley, 'The Irish in Modern Scotland'(Cork, 1947), p. 167.
2 E. J. T. Collins, Harvest Technology and Labour Supply, Nottingham University PhD Thesis, 1970, pp. 107-8.
3 George Bourne, 'Lucy Bettesworth' (1913), p. 140ff.
4 See Chapter 3, p. 49.
5 'Morning Chronicle', 26 Dec. 1849.
6 See pp. 79-80.
7 Richard Jefferies, 'Chronicles of the Hedges' (1879), Samuel J. Locker (ed.) (1948 edn), pp. 181-2.
8 Joan Thirsk, 'English Peasant Farming' (1957), p. 323, 'the Irish moved from south to north keeping pace with the ripening corn'.
9 Collins, Harvest Technology and Labour Supply, p. 137. Estimated numbers of Irish: 22,000 in 1810; between 63,000 and 75,000 at their peak between 1840 and 1850; dropping from approximately 45,000 to 33,000 between 1870 and 1880; but see Figure 5.1 above.
10 Jefferies, 'Chronicles of the Hedges', p. 93; Collins, Harvest Technology and Labour Supply, pp. 107-8.
11 E. J. T. Collins, Harvest Technology and Labour Supply in

Britain 1790-1870, 'Econ. Hist. Rev.', vol. XXII, no. 3 (1969) p. 472. By the mid-1860s the Irish element was 'nearly extinguished' in Middlesex, Lincolnshire, Yorkshire, Wiltshire and Sussex. But cf. 'Oxford Chronicle', 31 Aug. 1895, p. 6, col. 3. Farmers report 'a large influx of Irish Harvestmen' (S. Lincs).

12 'Ag. Gaz.' (1867), p. 888; Cheshire, Shropshire and Warwickshire, p. 890; Gloucestershire, p. 891; Essex.

13 Jefferies, 'Chronicles of the Hedges', p. 183; ibid., p. 93; Richard Jefferies, 'Nature near London' (1883), p. 83 (Surrey).

14 'Aylesbury News', 5 Sept. 1846. Though the Oxford and Rugby railway was in progress no shortage of labour was reported. In Berkshire, 'a more admirable wheat harvest has seldom been witnessed'.

15 Barbara Kerr, 'Bound to the Soil' (1968), p. 107. 'Irish migrants avoided Dorset and Wiltshire where they recognised a poverty equal to their own.' But note the source; figures for Irish-born from the only June census (1841) - a very dubious one.

16 'Oxford Chronicle', 5 Aug. 1854.

17 Oxford RO, DIL 1/e/2a, Farm Account Book.

18 F. G. Heath, 'British Rural Life and Labour' (1911), p. 148.

19 Ibid.

20 JOJ, 27 Sept. 1873.

21 OT, 17 Aug. 1889.

22 T. MacKay, 'The Reminiscences of Albert Pell' (1908), p. 144.

23 Flora Thompson, 'Larkrise to Candleford' (1962 edn), p. 258.

24 MacKay, 'Albert Pell', p. 144.

25 'Banbury Guardian', 10 Aug. 1848.

26 RLSC, 5 Aug. 1848.

27 Ibid., 29 July 1848.

28 Ibid., 19 Aug. 1848.

29 Ibid., 19 Aug. 1848; ibid., 26 Aug. 1848; ibid., 2 Sept. 1848; ibid., 14 Oct. 1848.

30 Ibid., 19 July 1848.

31 'Warwick Advertiser', 18 Aug. 1860.

32 Ibid., 7 July 1860; ibid., 28 July 1860.

33 Ibid., 25 Aug. 1860. Both occasions involved appearances at Warwick Petty Sessions.

34 RLSC, 21 Aug. 1880, two Irish arrested out of 14 causing a disturbance at Kenilworth; ibid., 14 Aug. 1860, four Irish out of 16, when a knife was produced on a farm outside Kenilworth.

35 Quoted in J. E. Handley, 'The Irish in Scotland 1798-1845' (Cork, 1943), p. 39.

36 Jefferies, 'Chronicles of the Hedges', p. 283.

37 'Warwick Advertiser', 25 Aug. 1860.

38 PP 1867-8 (4068) XVII, 'RC Employment... in Agriculture' App. pt. II, p. 530. But contrast male lust - RLSC, 5 Aug. 1848. For rape at Coughton, nr Alcester, ibid., 14 June 1848

- Irishman received 15 years transportation. For living conditions in Ireland, see E. E. Evans, 'Irish Folkways' (1957), pp. 86-7; '[families] lay together in one bed... visitors too slept with them... and all in the bare buff'. For Scotland, see Handley, 'Irish in Modern Scotland', pp. 185-6; 'housed in ... lofts, barns... and other outhouses. Beds... seldom provided... straw or hay placed on floor...'; and 'Oxford Chronicle', 16 July 1853, where a disastrous fire and loss of life was caused by Irish tramp haymakers leaving a box of lucifers behind in barn at Worley, nr Brentwood, Essex.

39 Jefferies, 'Nature near London', pp. 82-4.

40 'Banbury Guardian', 2 Nov. 1848 - two natives of the Emerald Isles charged with begging; Handley, 'Irish in Scotland', p. 45 - 'as our Androssan friends may term them'.

41 'Warwick Advertiser', 13 Oct. 1860, Leamington Petty Sessions, Wed. 10 Oct.

42 A. Redford, 'Labour Migration in England 1800-1850' (Manchester, 1926), p. 149. 'The Irish never secured a footing in the more highly skilled branches of agriculture.' But cf. 'Agricultural Statistics', 1900, p. 541. 'I was head ploughman on a farm in Lancashire for six years' (Drummin Fincell holders) - Irish harvesters.

43 MacKay, 'Albert Pell', p. 144.

44 'Warwick Advertiser', 9 Sept. 1848.

45 Cf. E. E. Evans, 'Irish Folkways', pp. 151-2; a scythe, a relatively new tool, 'what hay... made, cut with the [smooth-edged] grass or scythe hook'.

46 A. D. Hall, 'A Pilgrimage of British Farming' (1913), p. 75.

47 Handley, 'Irish in Modern Scotland', p. 173, 'the Irish labourer still finds it worthwhile to come'.

48 MacKay, 'Albert Pell', p. 145.

49 John Garwood, 'The Million-peopled City' (1853), p. 315.

50 Cf. John Denver, 'The Irish in Britain' (1892), p. 154, citing 'Stamford Mercury', Aug. 1850; 12,000 Irish passed through Liverpool to the Fens of Lincolnshire. But note; A growing and continuing practice of the London Irish. Multitudes of Irish 'have migrated to Kent for the hop-picking' (reported in 'Spectator' Sept. 1850); 800 from Poplar, chiefly Irish hop-picking (1891).

51 Handley, 'Irish in Modern Scotland', p. 169. Loss through emigration approximately 3 million.

52 Agricultural Statistics, 1900, p. 541. 'They start May 1st. Some return in September and some not till November' (meeting at Murrisk, 1892). 'We go to England between March and June and return in November or at Christmas' (meeting at Lesallagh, 1892). Statements made to A. Wilson Fox collecting evidence for the Royal Commission on Labour. Note: 'labour was so scarce... by 1860 Irish labourers were all employed all the year round in Lancashire and Cheshire' (J. D. Dent, Notes on the Census of 1861, JRASE, vol. XXV (1864), p. 230).

53 Jefferies, 'Chronicles of the Hedges', p. 183.

54 J. A. Jackson, 'The Irish in Britain' (1963), Appendix Table X, p. 193. From Londonderry and Portrush, 11 and 317; Belfast, 7,477; Warren Point, 1,740; Dundalk, 2,194; Drogheda 13,786; total 36,514 (1841). See also Handley, 'Irish in Modern Scotland', p. 172, Table II and figures for 1880, 1884, 1889, 1894, 1899 and 1904.

55 Agricultural Statistics, 1900, pp. 503-4; Heath, 'British Rural Life and Labour', pp. 148-9, for comments on this report and other facts.

6

THE ASSESSMENT OF HARVEST EARNINGS

> Harvesting... the one farm operation which has always created the exceptional demand for labour so as to fully extend the local labour market and even to require the recruitment of large numbers of casual and migrant workers.
>
> E. J. T. Collins, 'Sickle to Combine', pamphlet (University of Reading, 1968), p. 1.

The earnings of the rural worker at harvest time were the key to his survival in the nineteenth century. The day rate or the weekly rate which he was paid, taken at any decade in the nineteenth century, was completely inadequate to maintain a family at even subsistence level.[1] Harvest time meant all the family could earn and it was this family earning power, bolstered by piece work and harvest contracting, that made the vital difference. Time was the sanction that influenced the farmer. Time and the need of the farmer for labour made higher piece-work earnings available, and harvest contracting was for once a favourable bargaining area for the worker.

Apart from dairy farming all agriculture is a seasonal occupation. This was especially so in nineteenth-century England when the prosperity enjoyed between the repeal of the Corn Laws and the disastrous harvest of 1879 was based on the great increase in production of and demand for corn. Thus, there could be, on the one hand, high demand for labour in the summer months and, on the other, the apparently contradictory statement made that prior to 1870; 'there was in many country districts a superfluity of labour... a considerable proportion of the agricultural labourers returned as such [were] only in partial employment'.[2]

In the southern counties a growing rural population was set against a massive increase in arable production, making employment opportunities scarcer in winter with a greater demand for labour created in the harvest months. Even those fortunate enough to find regular winter work could not command much more than a subsistence wage in the slack seasonal periods, and in Berkshire, Buckinghamshire and Oxfordshire men were hired annually at Old Michaelmas (11 October) a slack time for labour. Thus, even by 1870, after almost two decades of profitable corn production, labourers received very little benefit in the form of any increase in the ordinary day rate,[3] despite the fact that prosperity was so well established.

An idea of the degree of prosperity is reflected in Table 6.1.

Table 6.1: Agricultural Hands and Product, United Kingdom (extract)

Year	Nos. of hands	Product (£ millions)	Product (£ per head)
1841	3,401,000	200	59
1851	3,519,000	220	63
1861	3,149,000	240	76
1871	2,808,000	250	89
1881	2,561,000	251	98

Source: M. G. Mulhall, 'Dictionary of Statistics' (1899), p. 15.[4]

It is not the purpose of this study to dispute J. H. Clapham's suggestion that 'broad divisions in earnings' between counties, especially between those in the north and south, 'correspond roughly with differences in labour efficiency'.[5] Indeed, E. H. Hunt has produced formidable advocates in James Caird, C. S. Read and George Culley so that 'we are left in no doubt that contemporaries believed that productivity varied substantially in different parts of the country'.[6] But precisely because normal wages were so low in counties such as Oxfordshire, Berkshire and Buckinghamshire, the opportunities for extra earnings that harvesting provided were crucially important to the rural proletariat.

Richard Jefferies, writing at this period, held the quite definite view that all labourers looked to 'the early summer haymaking and the corn harvest to supply, through extra wages then earned, the necessaries other than food - to pay rent, back debts, find shoe leather and so forth'.[7] It might be thought that because of extra harvest earnings together with their regular employment the hired men could improve their position and open a gap between their living standards and those of the day labourers who, according to W. C. Little's estimate in 1893, lost an average of 85 days work in a year from wet and wintry weather.[8] It is suggested here that this, in fact, did not occur, or only marginally. Only a full examination of harvest earnings can explain this seeming anomaly, and how the total payments made at harvest time affected the economic position of individual workers.

Disbursements made by farmers during the winter, compared with the summer months, recorded in contemporary account and labour books demonstrate how central to the agricultural worker's economic life harvest earnings were. As an example, the harvest began on 9 August on John Barford's farm at Fawcett, Oxon, in 1869. The wage bill for the last three weeks of the 'Harvest Month' amounted to £38 16s 0d. This did not include various payments for cutting at piece-work rates, which together totalled £16 1s 7d:

Dan Grd.	J. Payne 6 Ac Fagging Wheat	10s	£3 0 0
Sept 2	James Horn		£ s d
	2 Ac 0 2 Poles	13s	1 7 7
Sept 2	Joshua Allen & Brothers		
	10 Ac 1 rd 10 Poles Reaping	13s	6 14 0

The wages for the four weeks from 9 October, on the other hand, amounted to only £20 6s 0d, a difference of £34 11s 7d between payments to workers in the 'Harvest Month' (£54 17s 7d) and in a month of normal work (£20 6s 0d). The wages bill for the 'Harvest Month' beginning 24 August 1872 amounted to £50 14s 6d; by 16 November the previous month's wages had dropped to £22 6s 8d, a difference of £28 7s 10d. Interestingly, in 1884 with 'all the corn carried in the first fortnight but Beans very fine weather', the harvest wages dropped to £39 10s 9d, reflecting the effect of machinery on the amount paid to workers when reaping machines could operate in ideal conditions.

An 'Account for Labour' from 26 September 1863 till 15 October 1864 for a farm in the parish of Ripple, Tewkesbury, Glos, is presented graphically in Figure 6.1

Figure 6.1: Joseph Martin, Ripple, Tewkesbury, Glos: Account for Labour 26 September 1863 - 5 October 1964 (Total Labour Bill = £379 0s 10d)

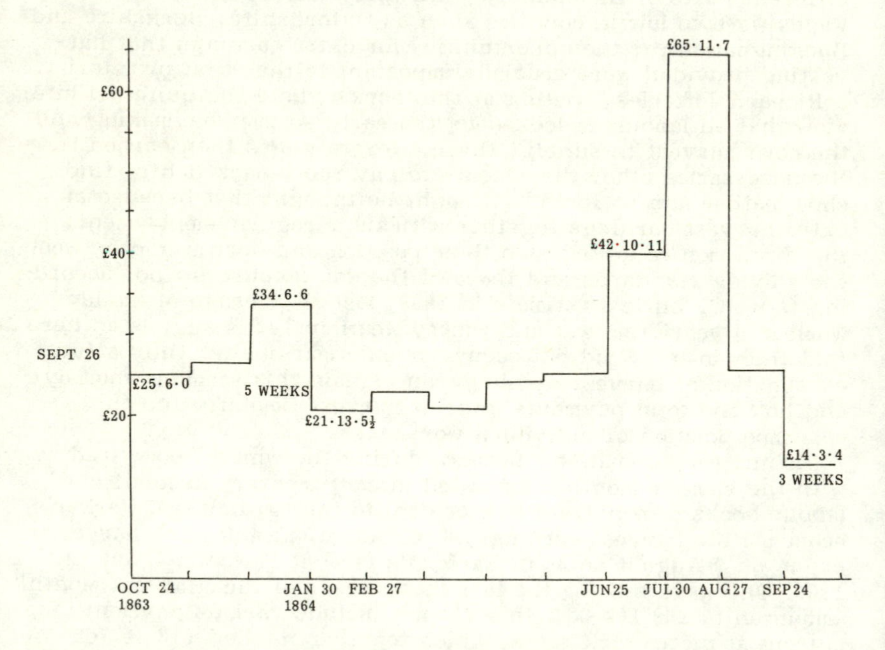

Source: Reading Farm Records, Glos. 2/2/2

The 'dead' month of the year, February, shows the lowest wage bill of £21 13s 5½d; for the 'Harvest Month' the wage bill is £65 11s 7d, presenting a difference of £40 5s 7d. July's wage bill was £42 10s 11d, significantly the month of the hay harvest with the proviso that payment was in this case for a five-week period.

On Wick Farm, Radley, Berks, the first week's labour bill in October 1853 was £8 16s 2d. In the first week in September 1853 - a week during the harvest - the wages paid amounted to £36 0s 5d. There was an even greater margin between a week after harvest and a week during harvest on the same farm in 1861. The wages paid for the week commencing 30 August amounted to £55 19s 11d. Three heavy payments for piece work are itemised:

		£	s	d
Aug 30	Fagging 43 acres wheat at 9s per acre	19	7	-
	Do 43 acres Barley at 9s per acre	19	7	-
	Pulling 20½ Acres beans ... 10s per acre	10	5	-

The wages, in contrast, for the week commencing October 1861 come to £8 8s 8d - a difference of £47 11s 3d. This was not exceptional at Wick Farm; in the harvest week 7 September 1860 a total of £42 1s 5d was paid for the piece-work fagging of barley and wheat out of the week's wage bill of £54 12s 8d. By September 1860, with the harvest over, the wages were as low as £6 14s 11d even with the inclusion of some back payments for turnip hoeing.

These examples give concrete support to Heath's remarks on harvest earnings in his 'British Rural Life and Labour': 'In corn districts, where the ordinary weekly wages are lowest, the emoluments of harvest are the most keenly appreciated'.[9] The problem of the economic condition of the agricultural worker in this period must be considered in the light of how informative contemporary reports are on the main recipients, the comparability of normal and harvest wages and systems of payments. Summaries of ordinary and harvest wages for the counties of Scotland, England and Ireland were published in 1850, 1860 and 1868 by the 'Gardeners' Chronicle and Agricultural Gazette'.[10]

It is useful, therefore, to examine in detail at least one of these reports since 'harvest wage data are notoriously scarce',[11] though bearing in mind that the farmers were the reporters. Of course agricultural earnings through most of the nineteenth century 'stubbornly refuse to be reduced to statistical form'.[12]

The return for Oxfordshire, Berkshire and Buckinghamshire for 1860 extracted from the 'Agricultural Gazette' shows the weekly wage of able-bodied ploughmen as follows:

Oxfordshire	12s to 14s (Eynsham)
Berkshire	10s and £1 at Michaelmas (Bornton)
	9s with house and garden free
	Potato ground free, beer and double wages

	in harvest (Hungerford)
	10s to 15s (Coleshill) and £1 in harvest (Streatley)
Buckinghamshire	11s (Claydon)
	13s to 14s (Bladwell)
	13s (Windsor)
	12s (Woolston)

And the wages of shepherds:

Oxfordshire	11s to 13s and 2s 6d a score for shearing (Eynsham)
Berkshire	10s and £1 at Michaelmas (Bornton)
(and 1s per double couple)[13]	9s with house and garden free, potato ground free, beer and double wages in harvest (Hungerford)
	11s to 14s house free (Coleshill)
	11s house free (Streatley)
Buckinghamshire	11s (Claydon)
	12s to 14s (Bladwell)
	12s (Windsor)
	12s (Woolston)[14]

The above wages applied to the annually employed, ostensibly an elite in terms of wages and conditions, certainly in terms of security of employment. The rates are apparently disparate but differences are inconsiderable when payments in kind approximating in value to extra in cash are considered, while relative hours of work and responsibility have to be taken into account. Since the rates could well be representative of the pay on larger and more ordered farms, many hired men's pay could have fallen below, while the opportunities for higher piece-work earnings for all hired men were generally restricted by the nature of their work. The possibility of extra earnings, then, were variously built into the annual hiring agreements (Bornton, Berks), specifically in cash and in kind (Hungerford, Berks), unspecified generally in Oxfordshire and Buckinghamshire.

Usually it is not the lack of knowledge on wage rates, but the presence of other circumstances which make interpretation of the agricultural workers' economic condition, even within county boundaries, difficult. Yet the rates themselves are always open to question. The generally higher rates in Buckinghamshire, for instance, are seriously put in doubt by the strike of 'Agricultural Labourers' at Gawcott in March 1867: 'They have worked all winter for 9s and 10s a week and now they demand 12s and 1s for Sunday work... their demand was refused and they were... paid off at the lower rate.'[15]

The weekly rates given for the 'ordinary wages of common labourers' in 1860 were:

Oxfordshire	10s to 12s (Eynsham)
Berkshire	10s (Bornton)
	8s to 9s with beer at certain times (Hungerford)
	8s to 10s (Coleshill)
	10s and £1 at Michaelmas (Streatley)
Buckinghamshire	10s (Claydon)
	10s to 12s (Bladwell)
	12s (Windsor)
	10s (Woolston)

The weekly rates for 'Harvest wages of common labourers' were:

Oxfordshire	10s to 15s with cottage 1s, beer 1s, and £1 in harvest (Eynsham)
Berkshire	10s with meal and drink (Bornton)
	15s with beer at certain times (Hungerford)
	12s with beer at certain times (Coleshill)
	no rate supplied (Streatley)
Buckinghamshire	16s (Claydon)
	20s with beer at certain times (Bladwell)
	24s (Windsor)
	26s (Woolston)

And the 'Ordinary [day rate] and harvest wages of women working in the field':

Oxfordshire	8d to 1s, no harvest rate supplied (Eynsham)
Berkshire	generally ranging from 7d to 10d. 1s during harvest
Buckinghamshire	generally 8d to 1s, no harvest rate supplied

The following are the rates supplied for piece work in hay time:

Price per acre of cutting clover and meadow:

Oxfordshire	3s 3d ;	3s 3d (Eynsham)
Berkshire	3s 6d ;	3s 9d (Bornton)
	2s 6d to 3s ;	3s to 3s 6d (Hungerford)
	2s 6d to 3s ;	3s 6d (Coleshill)
	2s 6d to 3s ;	3s 6d to 4s 6d (Streatley)
Buckinghamshire	2s 6d ;	4s (Claydon)
	3s 6d ;	4s (Bladwell)
	4s ;	5s (Windsor)
	2s 6d ;	3s 6d (Woolston)[16]

And at harvest:

Price per acre of cutting and tying wheat:

Oxfordshire	12s (Fynsham)
Berkshire	11s (Bornton)
	10s to 12s (Hungerford)

	10s (Coleshill, Streatley)
Buckinghamshire	12s (Claydon)
	10s to 15s (Bladwell)
	13s (Windsor)
	16s (Woolston)

And finally the charge for rent of cottages:

Oxfordshire	1s to 2s weekly (Eynsham)
Berkshire	78s (Bornton)
	50s to 70s (Hungerford)
	£2 to £5 (Coleshill)
	£2 to £3 (Streatley)
Buckinghamshire	52s (Claydon)
	78s to £6 (Bladwell)
	£5 (Windsor)
	52s (Woolston)[17]

From the above facts and figures it is shown that the extra earnings at harvest stemmed from an enhanced day rate, sometimes with a fixed payment at the end of harvest. The rate could be as much as double the ordinary day rate, i.e. 24s (Windsor, Berks), with a range of payments from the ordinary day rate with payment made in kind, i.e. 10s with meal and drink (Bornton, Berks), to the abnormally high rate of 26s without beer, 16s more than the ordinary day rate (Woolston, Bucks). Another main indeed the most considerable source of increased earnings was from the piece-work cutting of both hay and corn crops, the rates at hay time ranging from 2s 6d per acre (Hungerford, Coleshill, Streatley, Bucks; Claydon and Woolston, Bucks) to 4s per acre (Windsor, Berks) for cutting meadow hay, and from 10s per acre (Coleshill, Streatley, Berks) to 16s per acre (Woolston, Berks) for cutting and tying wheat.[18]

But it is clear from this data that important facts are not, or are only in part, revealed and affect the real wage rate. These were: factors determining individual earnings, e.g. category of work, length of hours, state of crop and so forth; the type of worker who directly benefited from different systems of payment; the variety of piece-work rates for crops other than wheat, e.g. barley, oats, beans, depending on variations in cutting methods (including wheat); the variety of earning opportunities for workers other than the able bodied, e.g. lower paid regular farm staff - lads, women and children, casual workers, and, most importantly, companies of harvesters and migrant harvesters, whether Irish or indigenous; and the existence of other payments both in cash and kind.

The main drawback of such reports is that no single set of rates could encapsulate the complexities involved in determining payments, especially at harvest time, which took into account the type of work, tools employed, availability of labour, type and state of crop, provisions of payment in kind and status of workers.

These reports must therefore be seen as frameworks indicating the range and variety of the wage structure above which earnings rarely rose but below which the payments received by many workers could fall.[19] They might well be seen as guides to farmers encouraging uniformity in payments in a complex wage structure rather than as evidence of definitive wage rates.[20] That there was a tendency towards uniformity in total earnings, it is suggested, was one of the most important social and economic factors in the lives of the agricultural labourer in the nineteenth century.

A much more revealing survey framed in terms of questions and answers put to farmers regarding harvest rates and practices was published by the 'Agricultural Gazette' in 1867. It discloses not only the involved nature of harvest work but shows how the relationship of tools used and crop requirements regulated actual payments at piece work. The answers are quoted at length because they show the division of labour and indicate the presence of different types of labour; that a change in hand-tool use had evolved; that a change over to machinery in the cutting process was still in the embryonic stage; that custom and change were not uniform in adjacent counties, local conditions being generally responsible for retaining these customs and for initiating any change.

The following are answers to questions relating to 'wages by the month or day during harvest time'.

> [In Berkshire:] From 18s to 22s per week; beer on carting days [J. Adnams]. From 2s 8d to 3s per day [W. Balstrode]. [In Buckinghamshire:] Reaping and mowing per acre, Carters, Shepherds and stock men, £4 4s and upwards for the month with beer [W. G. Duncan]. 3s a day [W. Smith]. £4 for the month or by the day 3s, with beer [J. K. Fowler]. [In Oxfordshire:] 2s 6d per day with an allowance of malt or beer [S. Druce]. All at piecework as much as possible. When carting commences the best men receive 18s per week [M. Savidge]. Generally 2s per day with beer; about a gallon and sometimes more, to a man, when at day work amongst corn [C. Wallis]. The men employed by the day during the whole time of harvest have from 20s to 25s extra and 2 bushels of malt each; those who have the chance of piecework, and are occasionally employed by the day, have 3s per day and an allowance for beer [G. Garne]. 3s 4d per day; no drink or extras.[21]

In these anwers the specific distinction between the cutting process and the work of gathering is made, payment at piece-work rates applying to the former. These can be summarised as: an increased day rate with beer (J. Adnams, Berks; J. K. Fowler, Bucks; S. Druce, G. Garne, Oxon). An increased day rate without beer (W. Bulstrode, Berks; M. Savidge, G. Garne, Oxon). Generally higher but in fact the equivalent of having beer supplied free, but benefiting non-drinkers. The harvest 'extra' either paid as a total payment: £4 4s and upwards for annually hired men (W. G. Duncan, Bucks) or in addition to ordinary day-rate pay; 20s to 25s for the 'Month' men, i.e. day men contracting for the harvest month (G. Garne) - all with free beer.

The following are the answers to questions relating to 'the prices for cutting Wheat, Barley and Oats per acre':

> [In Berkshire:] Wheat, cutting, tying up and shocking from 10s to 15s per acre; Oats and Barley if mowed, say 2s to 3s per acre without beer [J. Adnams]. Fagging Wheat from 11s to 16s per acre; Fagging barley generally about 12s per acre; mowing Barley from 3s to 5s per acre [W. Bolstrode]. [In Buckinghamshire:] Wheat from 9s upwards; Barley from 3s upwards; Oats from 7s 6d upwards [W. G. Duncan]. Wheat 6s to 16s; Barley 3s to 5s; Oats 3s per acre [W. Smith]. As crops are so variable so must the price be; a good upstanding crop of either Wheat or Oats 12s per acre; Barley, fagged and sheaved, 10s per acre; but layered crops vary from 14s to 25s per acre [J. K. Fowler]. [In Oxfordshire:] 10s to 15s per acre for Wheat; 8s to 10s for Barley, tied up in sheaves; 8s to 10s for Oats ditto; 2s 3d to 3s 6d if mowed and not tied - all exclusive of beer [S. Druce]. Most of the crops are now fagged at various prices from 10s to 14s per acre; if Barley is mown, 2s 6d [M. Savidge]. Always cut by the acre, from 8s to 18s per acre for the fagging or reaping. Fagging most general. A considerable quantity of the Barley is fagged, and nearly all the heavier crops of Oats. For mowing Barley or Oats, about 3s per acre. Sometimes more. Reaping machines are becoming thicker [G. Wallis]. From 8s to 14s per acre for fagging [G. Garne]. 10s for fagging Wheat, Barley and Oats; 5s for peas.[22]

A study of these answers resolves some of the complexities regarding variations in the piece-work payments. Generally a high rate was paid when piece-work involved not only cutting but tying and shocking the crop - wheat from 10s to 15s per acre (J. Adnams, Berks), a low rate for merely mowing and 2s 6d if barley was mown (M. Savidge, Oxon). Reaping with the slower reaping hook commanded a higher rate than fagging with the faster fagging or bagging hook - from 8s to 13s for the fagging or reaping (G. Wallis, Oxon). The state of the crop, heavy or light, also affected the price, 'as crops are so variable so must the price be... layered crops vary from 14s to 25s per acre' (J. K. Fowler, Bucks).

The following are answers to questions relating to 'the quantity of beer or cider given daily or per acre':

> [From Berkshire:] Corn is cut by the acre at a given price, without beer [J. Adnams]. From six to eight pints of beer per day when carrying according to the length of day; none when at taskwork [piece work] [W. Bulstrode]. [From Buckinghamshire:] 4 pints of ale and some small beer daily, for

the month's men [W. G. Duncan]. No beer for contract work, 4 pints per day for day work [W. Smith]. Four pints per day when not working by the piece [J. K. Fowler]. [From Oxfordshire:] The carting and stacking men only have an allowance of malt or beer [S. Druce]. Where cider is made the quantity consumed is enormous; on this estate no drink is given in part payment. The best men receive 20s for haytime and again 20s for the harvest; the rest of labourers - men, boys and women - at what they are worth [M. Savidge]. We give no beer or cider with the piecework [G. Wallis]. Without beer or cider [G. Garne].[23]

These answers can be generally summarised as beer provided for all work other than piece work (at task work, Berkshire); where no beer is provided on day work usually the rate is higher than normally enhanced day rates (cf. G. Garne).

The following are the answers to questions relating to 'the mode adopted of harvesting Wheat, Barley and Oats':

[From Berkshire:] Corn is all cut at task work and carried by day work; reaping machines very little used; nearly all corn cut with the fagging hook, except Barley, which used to be all mown; fagging this is however, becoming more general every year [W. Bulstrode]. [From Buckinghamshire:] Wheat is reaped fagged and mown according to circumstances, tied and set in shocks. Barley is mown and carried loose. Oats generally fagged and tied, or mown and tied. Beans are mown and cocked or fagged and tied according to length of straw and bulk of crop; a great many Bean crops this year cannot be tied [W. G. Duncan]. Some reaped, some mown, some cut by machine, and some fagged [W. Smith]. Fagging has now become prevalent... there are not many reaping machines in use in this neighbourhood. Farmers are looking out for the best class of reaper... none have been seen yet that will cut a layered crop perfectly [J. K. Fowler]. [From Oxfordshire:] The Wheat is all tied in sheaves, and latterly the Barley and Oats have also. All the harvest is done by a regular staff of men, who are employed summer and winter; being near a large village, some few small tradesmen turn out and assist in cutting the corn for two or three weeks during the busy season. Reaping machines are used but little at present in this neighbourhood [S. Druce]. Fagging has become very general in this part, and many good hands are to be found; the last wet harvest has made the farmers anxious to sheaf all their corn, if possible either by hand or machine; when properly done it will stand much wet and damage [M. Savidge]. Mostly by fagging. I tie up all I can. The few fields that are not laid will be cut with Samuelson's shearing reaper, and tied by the acre; but mine being mostly heath land, the corn grows frothy and rather weak in the straw, so that it is difficult to cut by machine [G. Garne].

> All corn sheared and carried by one-horse-carts; the pitching and ricking etc. all day work at 3s 4d.[24]

Besides the information regarding the type and extent of extra payments to regular workers contained in this survey, it indicated that others, i.e. small tradesmen [S. Druce[, women and boys found opportunities for extra earnings 'at what they are worth' [M. Savidge). Its considerable importance, however, is due to the fact that it suggests that machinery was still, in 1867, very much in the embryonic stage: 'it is still difficult to cut by machinery' (G. Garne); 'used little at present' (S. Druce); 'none has been seen yet that will cut a layered crop perfectly' (J. K. Fowler). The main cutting tool in Oxfordshire and Berkshire, though less dominant in Buckinghamshire at this period, is shown to be the fagging hook. But the use of reaping hook and scythe in conjunction with the fagging hook indicates the range of different crop circumstance that still made each an economic cutting instrument.[25] By implication there was, therefore, as yet no immediate threat to piece work, the most profitable method of taking work at harvest time. On analysis, however, as in the case of day rates, apparently widely different rates did not produce any wide variation in piece-work earnings, since the rate was calculated on the speed of the cutting tool in use and the state of the crop, and only exceptional skill and endurance involving longer hours of work established any wide difference in individual earning capacity. If W. G. Heath could say with truth that 'There is an anormous difference in the piece work capacity of different men' and describe a Somerset man of his acquaintance as 'strong and well built, with sinews like iron, [so that] no one could touch him for the work he did',[26] such a man must be set against the 'widow' man who, having no 'followers' to make bands and tie and shock the corn he cut, fell behind his more fortunate companions in the harvest field.[27] The widespread practice of collective piece work undertaken by 'companies' of harvesters who shared the reward of joint effort also suggests that such men were never prototypes of agricultural man but rather the necessary archetypal figures of any social group.

It is their limitation that such reports mask the existence of such workers as 'followers', the women and children who, though not paid by farmers, affected the profitability of piece work, and the fact that labourers, by working in 'companies' were stimulated to greater effort yet equalised out their individual earnings. But it is implicit in this report of 1867 that there was no immediate danger of any considerable reduction in the opportunities for piece work or a threat to the earnings of those workers who relied so heavily on this kind of harvest work involving the various hand-cutting tools. There is also the indication that continued technological change was to remain for some time centred on a switch in hand tools rather than in any rapid introduction of machinery.[28] Indeed, H. Rider Haggard was still in 1898 echoing the same objections to the reaping machine as those expressed in 1867:

> [T]he price of... a reaper... is... a good deal of money [£26] to expend on one article... [it is] a mistake to suppose... a reaper will cut corn in every case and every season... [it] does more harm than good [if corn is] 'badly laid and twisted'... [the] treading of the horses is too destructive of [barley].[29]

NOTES

1 F. G. Heath, 'The English Peasantry' (1874), p. 187. 'Nothing short of combination would effect any improvement in the deplorable condition of the peasantry' (quoting Canon Girdlestone at British Association meeting, Norwich, 1868).
2 PP 1906 (CD 3273) XCVI, Report on the Decline in the Agricultural Population of Great Britain 1881-1906, p. 9.
3 Heath, 'The English Peasantry', for wages in Berks in 1870: 10s-11s, pauperism 6.5%; in Bucks, 11s-13s, pauperism 6.3% (p. 9); in Oxon, 10s-11s, pauperism 4.7% (p. 20). Compare wages in the north: Cumberland and Westmorland, 15s-18s, pauperism, 3.8%; Derbyshire, 14s-17s, pauperism 2.4% (p. 11); Durham, 15s-18s, pauperism 3.5% (p. 13).
4 PP 1906 (CD 3273) XCVI, p. 10.
5 J. H. Clapham, 'An Economic History of Modern Britain', vol. IV (Cambridge 1951), p. 97, though modified by 'Yet differences in efficiency can hardly explain... the 14s 6d of Oxfordshire - the lowest figure in England - the 16s 4d of Buckingham...' (figures for 1902).
6 E. H. Hunt, Labour Productivity in English Agriculture 1850-1914, 'Econ. Hist. Rev.', 2nd Ser., vol. XX, no. 2 (1967), pp. 281-3.
7 Richard Jefferies, 'Chronicles of the Hedges' (1879), Samuel J. Looker (ed.) (1948 edn), p. 93.
8 PP 1893-4 XXVII pt. II, Abstract of RC on Agricultural Interests 1879-1882, p. 301.
9 F. G. Heath, 'British Rural Life and Labour' (1911), p. 31.
10 'Ag. Gaz.', 27 Apr. 1850, pp. 266-7, 30 Apr. 1860, pp. 392-3, 16 May 1868, pp. 526-7.
11 E. J. T. Collins, Harvest Technology and Labour Supply in Britain 1790-1870, 'Econ. Hist. Rev.', 2nd Ser., vol. XXII, no. 3 (1969), p. 464 footnote.
12 E. P. Thompson, 'The Making of the English Working Class' (1968), p. 235; the complications of averages, pp. 233-7.
13 Cf. Labour Book entry, Badcock, Radley, Berks, 7 Oct. 1836; James Bennett, 1s each for 14 Double Couples - 14s. BER 13/5/2, 1s for twin lambs reared.
14 'Ag. Gaz.', 30 Apr. 1860, pp. 392-3.
15 Ibid., 30 March 1867, p. 327.
16 Ibid., 30 Apr. 1860, pp. 392-3.
17 For the relevance of cottage rent, see below, pp. 105-6.
18 'Ag. Gaz.', 30 Apr. 1860, pp. 392-3.

19 'Labourers' Union Chronicle', 9 Aug. 1873, p. 6, col. 1. Many men in the Aylesbury district worked in hay harvest without overtime payment.
20 Ibid., 14 June 1873, p. 1, col. 2. 'Farmers... arranged at the church gate on Sundays... how little they would give for mowing and reaping!'
21 'Ag. Gaz.', 24 Aug. 1867, pp. 890-1.
22 Ibid.
23 Ibid.
24 Ibid.
25 See above, p. 98.
26 Heath, 'The English Peasantry', p. 26.
27 See Chapter 9.
29 See Collins, Harvest Technology and Labour Supply, p. 454ff for a detailed analysis of 'Labour saving methods... developed within and alongside an existing framework of traditional hand tool techniques at harvest time.'
29 H. Rider Haggard, 'A Farming Year' (1899), p. 274.

7

DISTRIBUTION AND METHODS OF HARVEST PAYMENTS

> It is only necessary to compare the weekly budgets with the weekly earnings to realise that the large majority of labourers earn but a bare subsistence, and are unable to save anything for their old age, or for times when they are out of work. An immense number of them live in a chronic state of debt and anxiety, and depend to a lamentable extent on charity. Their cottages are bad and often contain a minimum of furniture. It is very difficult for them to get milk for their children, and the supply of good water is in many districts deficient. For six days of the week they live on vegetables, bacon and bread, and on Sunday the change is more often to pork than to beef or mutton.
>
> PP 1893-4 XXXV, RC on Labour; the Agricultural Labourer, p. 200

That the circumstances of the agricultural worker in the nineteenth century were poor is not in dispute; that harvesting was the central focus of their survival and decided the state and degree of their poverty is a matter of fact; how they remained fixed in poverty is a matter of interpretation. First, two types of agricultural worker have to be distinguished: those in the northern counties close to urban industrialisation, with the consequent upward pull to wages, and where the upstanding wage system prevailed (that is a wage going on notwithstanding the weather or sickness),[1] and those in the purely agricultural eastern and southern districts (Caird's line) with low basic wages subject to the economic pull of overpopulation in winter towards a subsistence level. Within these counties, the concern of this study, the agricultural workers can be placed in three distinct categories: First, the annually hired labourers, the carters, shepherds and cattlemen, direct tenants of their masters or inhabitants of closed villages, who exchanged freedom for a dubious security, and who were engaged for their skills but who were paid in the event for longer hours and a commitment to work on Sundays; second, the inhabitants of the open villages who could find work near or far, preferring an uncertain freedom to subservience; and, finally, those other inhabitants of open villages who through lack of skills or initiative could neither obtain the security of annual employment nor make full use of the opportunity to obtain work at the higher rates prevailing during the harvest period.

Harvest time therefore had three aspects in the economic life of the different kinds of agricultural worker. The closed villagers, enjoying some security, looked to the completion of harvest since most of the economic advantages of harvesting were already structured into their annual agreements. The open villagers with skills and initiative, however, looked to the commencement of harvest. This was the time of certain employment. Work was not only freely available but the rate of pay could be 'bargained'.[2] Unlike the hired men tied to a single farm or limited by the nature of their work from obtaining high earnings from piece work, they were limited only by their strength and endurance or their reluctance to travel. For the rest of the open villagers the harvest meant a brief period of work, a temporary release from pauperisation. In the summer they constituted an important pool of harvest labour - for the rest of the year they were condemned as a drain on rural society though their poverty had in part been created by the needs of the new commercial agriculture. Only at harvest time were the issues wages and not merely work.

Evidence of this relationship of earnings to these three categories of workers can be found in the labour books of the period and in conjunction with the contemporary 'blue books' all the various methods of payment in cash and in kind can be identified. The 'Agricultural Gazette' reports were entitled the 'Value of Agricultural Labour', but the farm accounts show how misleading such a generalisation was, for lads and boys hired at much lower rates were very much part of the regular labour force, while women and children, especially at harvest time, were also an essential part of farm labour cost and in supplementing the labourers' wages increased the overall family income.

The complications and details of harvest payments were indeed many and various; some were well established before the nineteenth century; generally a single method was not adopted by itself. Such a single method, however, was in operation on the Ditchley Park Estate in North-West Oxfordshire during the eighteenth century. Variously termed 'allowance for Harvest', 'advance for Harvest' or 'Harvest Extra', a lump sum was paid at the end of harvest, with the normal day rate remaining unchanged for this four-week period.[3] In the summer of 1756, 'the wettest... in the memory of man',[4] the 'extra' at Ditchley amounted to 12s paid on 25 September; the ordinary day rate was 5s per week so that for every day worked each man received 1s 4d instead of his usual 10d. In the summer of 1765, 'remarkable for its heat and dryness',[5] the harvest was over by midday Saturday 31 August.[6] Though the ordinary rate had risen to 6s per week, the harvest 'extra' had been reduced to 8s. Comparing two entries for the same worker, one in the last week of harvest in 1756,

20 to the 25 September 1756	Days	£	s	d
Samuel Harding	6	0	5	0
Do allowance for Harvest	24	0	12	0

with the comparable week in 1765,

26 to the 31 of August 1765	Days	£	s	d
Samuel Harding	5½	0	5	6
Do allowance four weeks			7	10
Harvest extra				

it might appear at first glance that wages had fallen. In fact the day labourer, provided he was in full employment, could expect a year-round increase of £2 8s though his total earnings at harvest time remained the same for a full four-weeks work (1756 - four weeks at harvest at 5s per week plus 12s extra equivalent to 1765 - 4 weeks at 6s per week plus 8s extra). While the wage rates were back in 1770 to the 1756 figures (5s ordinary weekly rate, 12s harvest 'extra'), the simplicity of the 'extra' to cover payment at harvest was to remain at Ditchley until 1801.[7] Then different methods of payment applying to different categories of workers are clearly shown in the following extract from the Ditchley Park Estate accounts for the last harvest week of 1801;

1801 The Farm Labourers Bill No. 3 extra Allowance for Harvest included September 6

	Days		£	s	d
Paid Jno. Harris	... 26	at 20d - £2.3.4 extra £2 ..	4	3	4
Geo. Hichols	... 19½	.... at 3s per day	2	10	6
Jno. Castle	... 24	....	1	12	0
Jas. Gubings	... 16½	at 3s	2	9	6
Tho. Gubings	... 23½	at 3s	3	10	6
Wm. Kinch	... 23½	at 3s	3	10	6
Jno. Bolton	... 23½	at 3s	3	10	6
Wm. Field	... 23½	at 3s	3	10	6
Jno. Day	... 26	at 20d - £2.3.4 extra £2..	4	3	4
Do for 5 Sundays..................................				5	0
Hen. Hall 26 and 5 Sundays..........................			2	3	0
John Baylis	.. 26	...... extra 4s		17	0
Tho. Collings	... 26	...... extra 4s		17	0
Ben Sims	... 26	...... extra 4s		17	0
Wm. Padbury	... 24	....		14	0
Wm. Castle	... 18	at 3s	2	14	0
Chas. Mander	... 4	....		3	0
Geo. Nichols, Jas. Gubings, Tho. Gubings, W. Kinch, Jno. Bolton & Wm. Field for Mowing 100a 30r of Oats & Barley at 20d per Acre..............................			8	7	11
			46	6	7[8]

Here are examples of hired men earning no more than their normal higher rate of 10s per week but paid an extra £2; day labourers receiving enhanced day rates of 3s per day (double the normal 9s per week) and six of them sharing £8 7s 11d for 'contract work' as a 'company'; the harvest extra of 4s paid to lads working at their normal rate of 3s per week; and, lastly, those labourers who presumably were paid 'what they were worth'. Significantly in this one week the greatest extra earnings were those of some of the day workers:

Tho. Gubings	23½ at 3s	£3. 10. 6
(share of the contract work of mowing)		£1. 7. 0
total earnings		£4. 17. 6

This can be compared with one of the hired men's earnings: 'Jno. Harris... 26 days... extra £2... £4. 3. 4.'

Three-quarters of a century later the harvest 'extra' was no longer paid on the Ditchley Estate, but the pattern of higher harvest earnings for the insecure day men and lower harvest earnings for the hired men continued. In the harvest month of 1878 the hired men on the Fulwell Farm were paid an enhanced day rate, 4s 2d per day to the carter, 3s 4d per day to the under carter and the shepherd. The day labourers received the same enhanced day rate of 3s 4d but some had the opportunity of piece work. Two day men, Joseph Shepherd and John Rook, spent 21 days at harvest work, thatching ricks and working at Lodge Farm (the other farm on the estate) for which they received £3 10s each. For their piece work during the harvest they shared £15 7s 0d according to an entry in the farm Labour Book which reads:

Shepherd and Rook			
Mowing & Tying	10 a. 1r. 18p.	Oats at 10s	£ 5. 3. 7½
Tying	20 a. 0r. 31p.	Wheat and Oats at 6s	£ 5. 0. 11½
do	17 a. 0r. 12p.	Beans at 6s	£ 5. 2. 5
		Total	£ 15. 7. 0

Another day man, Thomas Fowler, spent 19 days on similar work earning £3 3s 4d; in addition he received £7 2s 6d for piece work:

Mowing & Tying	3 a. 3r. 8p.	Oats at 10s	£ 1 18 5
Tying	10 a. 3r. 8p.	Oats & Wheat at 5s	£ 2 14 0
do	8 a. 3r. 33p.	Beans at 6s	£ 2 13 9
			£ 7 6 2

In contrast to these individual earnings the carter received £5 for his 24 days work in the harvest, £6 3s 6d less than Shepherd and Rook, £5 9s 6d less than Thomas Fowler.[9]

The total extra earnings at harvest time were obtained through several methods of payment: Michaelmas money, an enhanced day rate, piece work, the harvest 'extra', an harvest agreement or contract, beer money, or indirectly through payments in kind, e.g. provision of beer or cider, sometimes the equivalent in malt and hops, sometimes meals. However, what decided the individually accumulated earnings was the opportunity to benefit from these payments because some were mutually exclusive. In fact, what also decided the final gain was the distribution of these payments in relation to individual conditions of employment. Arthur Wilson Fox, an assistant commissioner, rather masks any distinction in conditions in the southern counties, when at the turn of the century (1903) he contrasted the position of labourers in the north and south:

> in the high wage counties the labourers' income whether paid in cash or kind is more evenly distributed over the year, while in low wage counties which are the arable ones, the labourers' weekly wages are increased at irregular intervals according as the season enables him to work at piece work or harvest.[10]

This irregularity, however, only partially applied to the annually hired labourers (one-third of those employed),[11] though their lower normal wage rates compared with those in the north still made extra harvest earnings vitally important. Bowley is also somewhat misleading when examining agricultural earnings four years earlier, for he was content to revert to and refurbish Arthur Young's century old conclusion that:

> total earnings for a whole year amounted to 20% more than the normal rate arrived at by stating separately the rates for corn harvest, for hay harvest and for winter including a valuation of all food and drink given by the employer assuming that the harvest rate was paid for five weeks, the hay rate for six weeks and the winter rate for the remaining 41 weeks.[12]

But this indiscriminate apportioning of the various opportunities for extra earnings to all workers was no longer relevant. By the middle years of the century a more complicated wage structure had developed with the emergence of larger scale commercial farming. An almost precise division had occurred between hired workers and the increasing numbers of irregularly employed day labourers. There was also to be a further complicating factor - the introduction of machinery in the cutting process.

To obtain a clearer picture, therefore, the opportunities for hired workers and day labourers must be evaluated separately. For the hired labourers these were as follows.

Michaelmas Money ('Over' Money) and Hiring 'Earnest'. The amount varied according to age and experience. It was not exclusively a harvest payment but contained an element for skill and for longer hours worked before 'dawn' and after 'dusk', the traditional span of a working day.[13] Generally, the higher the payment the higher the cottage rental; rent indeed was often paid only after the receipt of Michaelmas money. The hiring 'earnest', a small, varying sum was paid sealing the agreement - increasing towards the end of the century (basically 1s). Sometimes Michaelmas money was the only extra payment; on the other hand the following examples explain the difficulties of generalisation even for the same farm. These agreements were made at Wick Farm, Radley, Berks on 6 October 1838:

> Hired John Purbrick at 6s the first half year and 7s the last and £3 at Michaelmas. Hired Christopher Radbone at 6s the first half year and 7s the last and £2 at Michaelmas if goes on well till Michaelmas £1 more. Two lads hired at 3s (3s 6d) and 2s 6d (3s) with £1 and 10s respectively at Michaelmas. All paid 1s earnest sealing the agreement.

No other payments were made during harvest.[14] At Manor Farm, Aston, Oxon, five different payments of 'over' money were made: £1 and 2s 6d 'Earnest'; £2 and 1s 'Earnest'; £2 no 'Earnest'; £1 10s and 1s 'Earnest'; £1 and 1s 'Earnest'; but other earnings came from piece work and so forth during harvest.[15]

An Enhanced Day Rate. The amount varied; generally up to double the normal day rate related to the harder work and to the extra length of the working day and status of workers.[16] Variations in the rate are in evidence on the Ditchley Estate in 1878 while on the Fulwell Farm the rate ranged from 4s 2d per day paid to the carter to 1s paid to a plough boy. Nine men were paid 3s 4d per day, 2 men 3s per day, a casual labourer 2s 6d, another 2s 2d a day, 3 women 2s per day, 2 other plough boys 1s 2d and 1s respectively.[17] At the Lodge Farm in 1879 the following day rates were increased at harvest time: 2s 4d to 3s; 2s 2d to 3s; 1s to 1s 6d; 10d to 1s 3d; but the farm foreman's rate of 4s remained the same.[18] In the fortnight ending 10 October the harvest work for five men, a lad and a boy variously consisted of pitching wheat, wheat carting, straw carting, mowing barley, building ricks, helping on ricks, leading 'forist' (the lead horse in the waggon team) and general harvest work. They were, however, paid 'beer money': £1 each to the men and 10s each to the lad and boy for the harvest month. On a Fawcett farm, Oxon, up to double the day rate was paid with alternative piece work though this was somewhat curtailed by the employment of two 'harvest companies':

> Sep 2 1869 Joshua Allen and Brothers 10ac 1rd 10 poles Reaping at 13s per acre - £6.14s.
> Mark R. South & Company 4ac 2rd 15 poles at 13s - £2.19.8: day work 18s in all £3.17.8.

This was the practice over a period from 1865 to 1884; no Michaelmas money was paid.[19]

Piece Work ('By the Great' 'Task Work'). This was limited by ordinary day-to-day commitments; some piece-work cutting of the crops could be fitted in;[20] generally the harvest work was organised so that carters, shepherds and cattlemen were engaged mainly in carting and stacking operations for which piece work was not the customary method of payment. Piece work was therefore less profitable for the hired men because it was the extended day, starting work early and finishing late, that made this method of working the most remunerative.[21]

The Harvest 'Extra'. This was a lump sum paid for the weeks of harvest taken to cover a period of four weeks when the ordinary day rate continued to be paid. This method precluded piece working or an increase in day rate but usually equalled an enhanced day rate paid over the harvest period. Payment of Michaelmas money was also not precluded but this depended on the annual hiring agreement. Again it is difficult to generalise. At Wick Farm, Radley, Berks on 14 October 1861, William East was hired as a carter 'to live in the Gooseacre Cottage, to have 11s per week, £2.10s at Michaelmas, 10s extra for the harvest... to pay 1s per week rent' (1s 'earnest'paid). It is to be noted that the rent was 2s more than the Michaelmas money and no other extras were given.[22]

Overtime. An hourly rate paid after working hours when these were paid at the ordinary day rates. Relatively unknown until the claim that a working day should consist of a definite number of hours and not the loose concept of from light to dark. Demands for higher ordinary wages and/or shorter working hours in the late 1860s made it necessary to define more clearly the working day.[23] As an alternative to an enhanced day rate it may have been less advantageous to labourers because it was essentially a time rate; the benefit, however, could come outside the harvest period because a definite claim to payment for extra work after normally accepted hours could be made. The following collection of answers from Oxfordshire farmers, published by the assistant commissioner Andrew Doyle in 1881, shows this development towards a more uniformly structured day, though there were still variations in hours.

What are the ordinary hours of labour (a) In Summer (b) In Winter?

(a) 7 a.m. to 5 p.m. Any time longer in hay and harvest.
(b) 7 a.m. to 5 p.m. or from light to dark. 15 years ago time was from 6 a.m. to 6 p.m.
(a) & (b) From 7 to 5 except hay and harvest.
(a) & (b) Labourers from 7 until 5 o'clock.
(a) From 6 to 6.
(b) From 7 to 5.
(a) Nine hours.
(b) Eight hours.

(a) 6 a.m. to 6 p.m.
(b) 7.30 a.m. to 4.30 p.m. according to light.
(a) Nine hours per day.
(b) Nominally the same.
(a) 7 a.m. to 5 p.m.
(b) Same.
(a) From 6.30 to 5.30.
(b) From 7 to 5.
(a) 7 to 6; harvest, 6 to 8; hay harvest, 7 to 8.
(b) According to length of day, short day's work from light to dark, or about seven and a half to eight hours.
(a) Shepherds, carters etc. 6 to 6. Day Labourers 7 to 5, except haytime and harvest.
(b) Same as above, except in shortest days.
(a) 7 a.m. to 5 p.m. One and a quarter hours for meals.
(b) The same. 7 till 6, after that paid extra.[24]

Beer Money. Beer was usually supplied free to all day workers but generally not to piece workers; if the piece work was taken at a lower rate, beer was sometimes provided. If no beer was supplied a cash equivalent, usually £1, was paid at the end of the harvest period, but never to piece workers.

Harvest Home Supper (Horkey in East Anglia and the Fens).[25] This supper provided by the farmer for the workers and their families to celebrate the completion of harvest. After the 1860s the practice of commuting the harvest supper into a money payment was gradually introduced and the sum of 2s 6d was paid to each worker (pro rata for women and lads, usually 1s). The provision for harvest supper at Hill Farm, Stokenchurch, Bucks, cost £1 6s in 1850.[26] At Fulwell Farm on the Ditchley Estate in September at the end of harvest 1878 eight men and two plough boys received £1 allowance each for beer and 2s 6d harvest home money, two day men received £1 beer allowance and 6d harvest home money each, one plough boy 8s, two others 7s each, three women 1s harvest home money only, and two casual workers 16s 6d and 14s 6d respectively.[27]

Payments in Kind. This consisted of free drink usually supplied only to day workers whether paid at day rates or at enhanced day rates; sometimes to piece workers when the piece rate was lower. Free drink and meals were sometimes the only payment made at hay time when working by the day. The cost of such perquisites on a farm at Holton, Oxon, was as follows: beer, tea and provisions supplied during hay making, corn harvest and for the harvest home supper, £18 13s 6d (1886); £17 11s 6d (1887); £11 14s 6d (1888); £10 5s 6d (1889); £10 12s 6d (1892), and £13 17s 9d (1893).[28] The opportunities for the day labourers were confined to an enhanced day rate, piece work and contract work. Their advantage lay in being able to switch from working by the day to taking piece work without being hindered by other commitments. At piece work not only could they work as many hours as they wished, to complete their work quickly, but their opportunities were not confined to a single farm. They could seek piece work on other farms and were not bound to take the less profitable day work even though the rate was greater than at normal times. Organised into bands they could bargain with the farmers for a profitable rate, especially when labour was in short supply. If a rate was unacceptable a refusal to accept the price offered could not incur possible unpleasant consequences which might have arisen had they been hired men. Most importantly, they could increase the profitability of piece work by enlisting family help.

The following payments made at Manor Farm, Aston, Oxon, in September 1883 indicate the superior piece-work earnings of day men compared with those of hired men, bearing in mind that piece-work opportunities were the most profitable:

	£	s	d
Hired man [£1 'over' money, 2s 6d earnest - plus other harvest pay]			
Geo. Luckett cutting and tying corn as per a/c	£3	18	1½
" " " beans 2r.19p. at 11s		7	1
" " tying corn as per a/c deducting 2s8½d for loose corn & carting and carrying (8s4¾d less 2s8d)		5	8½
Total	£4	10	11
Day labourers [plus other piece work or work at enhanced day rates]			
D. Baston cutting corn as per a/c	£7	19	4
" " beans 1a.0r.24p.		13	0
	£8	12	4
D. Ball cutting corn as per a/c	£5	9	0½
" " beans 1a.0r.28p. at 11s		13	4
	£6	2	4½
W. Harris cutting and tying corn as per a/c	£4	18	8½
" " beans 3r.1p. at 11s		8	4
	£5	7	0½
Chas. Clinch cutting and tying corn as per a/c	£4	8	1½
" " beans 1a.1r.9p.		14	6
	£5	2	7½
Mrs. Wearing cutting and tying corn as per a/c	£3	14	3½ [29]

An interesting point about the payments made to Geo. Luckett is that he was penalised for careless work and made to pay for the extra labour needed to collect the loose corn left untied. It may be noted, too, that Mrs Wearing's earnings were quite considerable. This could be so because piece work was the only method of payment where rates were not related to the sex or age of workers; indeed, paid individually, skilled women reapers using the light reaping hook, though not recorded statistically, could be of great help to the family economy.[30]

It might be said that only at harvest could there be free exchange between masters and men; despite the rigid hierarchy of rural society there was a measure of equality - a mutual immediate need - for at this period the farmer was dependent on the positive co-operation of all workers, to cope with all the uncertainties attending harvest work. The mood and attitude is perhaps best demonstrated in the making and carrying out of harvest contracts. A mutual trust once agreement was reached on price and work to be performed was essential - for once the labourers were not being treated as mere servants subjected to a continuing stream of orders but left to complete a complex series of tasks and skilled work using their own initiative. There is a sense of this transformation in the labourer's individual status in this description of the making of such agreements; Mark Rushton born at Larkfield on the Suffolk-Essex border in 1861 recalled the way these harvest agreements were made:[31]

> We were allus hired by the week, except at harvest. Then it was piece-wukk. I dessay you've heard of the 'lord', as we used to call 'im? Sometimes he was the horseman at the farm, but he might be anybody. His job was to act as a sort of foreman to the team of reapers - there was often as many as ten or a dozen of us - and he looked after the hours and wages and such like. He set the pace too. His first man was sometimes called the 'lady'. Well, when harvest was gettin' close, the 'lord' 'ld call his team together and goo an' argue it out with the farmer. They'd run over all the fields that had got to be harvested and wukk it out at so much the acre. If same as there was a field badly laid with the weather, of course the 'lord' would ask a higher price for that. 'Now there's Penny Fields' he'd say - or maybe Gilbert's Field - or whatever it was; 'that's laid somethin' terrible', he'd say. 'What about that, farmer?' And when the price was named he would talk it over with his team to see whether they'd agree. The argument was washed down with plenty of beer, like as not drunk out of little ol' bullocks' horns; and when it was all finished, and the price accepted all round, 'Now I'll bind you', the farmer 'ld say, and give each man a shilling.[32]

Arthur Randell reconstructs the making of a harvest contract and shows the peculiar nature of the relationship between master and men at this one period in the farming year:

> It was quite a business when the harvest men met the farmer each year to fix the price per acre for tying, shocking and carting... Often they would argue for as much as half a day but in the end they always came to some agreement and then the farmer would send for some beer to seal the bargain and a start would be made on the work.[33]

Though the change in attitude was important, it must not be exaggerated. A new sense of responsibility was allowed to be adopted by the labourers but in no sense did complete control pass out of the farmers' hands. Extracts from the two following agreements indicate the extent and nature of this control. On a Suffolk farm in the 1870s six men formed a company to take all the harvest work including the hoeing of the turnip crop. They agreed to

> cut and secure... all the corn grown on the farm in a workmanlike manner to my satisfaction, make bottoms of stacks; cover up when required, hoe the turnips twice and turn or lift barley once; turn the peas once - each man to find a gaveller [usually a woman - to gavel meant to rake into rows]. Another stipulation was that 'should any man lose any time thro' sickness he is to throw back 2d a day to the Company and receive account at Harvest. Should any man lose any time thro' drunkenness, he is to forfeit 5d to the Company.[34]

On 27 July 1847 detailed instructions were written into the harvest agreement when Poole's company of nine men undertook the complete harvest on an Essex farm at Paglesham with different stipulations according to the various crops, amounting in all to 108 acres. Wheat they were 'to reap, bind in small sheaves, trave and cart'; barley, 'to mow, gather, cart and due roke'; oats, 'to mow, gather clean and cart'; beans, 'to cut with hooks [fagging hooks?] tie trave and cart'. They also agreed to

> have three day's work, each man at 5s per day, to do anything I please, from 4 o'clock in the morning till 9 o'clock at night. For the sum of fifty four pounds, and to have 5 quarts of beer each man per day, when fine weather and at work, for the first fortnight and after that 4 quarts per day each man. You also engage with me Thomas Stebbing to do all the aforesaid harvest work in a good, workmanlike manner, pick the corn up clean, cut the stubble close (not higher than 8 inches), make small sheaves, bind them tight, set the tucks inwards and keep the sheaves set up till the field is finished cutting, but not to bind or cut when wet. And, each when at cart, you are all to help clean and bate the horses well. And each man shall forfeit, for every day he loses through drinking or getting drunk, six shillings per day and 3s for half a day so lost. And should any man

fall ill and obliged to lose time, another man will be put in his place during his illness.
As witness our hands this 27th day of July 1847,
Thomas Stebbing, Master, John Pool Foreman [and 8 others].[35]

This was indeed a very tight control and awareness of the details of harvest work. The expression 'trave' is an example. This did not merely mean 'shock' but designated the number of sheaves in the shocks, because although there were county variations in forming shocks, they also varied as to the number of sheaves.[36] A further curious distinction can also be made for in Oxfordshire the 'thrave' was used as a piece-work measure. At Wardington in 1867 cutting 'was done by the acre and also by the thrave or dozen'. The sheaves had to measure 22 to 24 inches round the band. There was also an old measure of 36 inches,[37] but though this was thought 'too big' at Wardington, it was still the practice at Epwell at this period to fag and tie by the 'thrave' which here was taken to be 24 sheaves. Paid at 4d per thrave piece-work earnings could amount to 5s a day.[38]

More than 40 years later the classic form of the 'Paglesham' agreement can be seen as contained in two agreements, though drawn up in another county, and though one of the instructions concerned the use of machinery. They relate jointly to Underwood Hall and Partridge Hall Farms, Borough Green, Cambridgeshire. In 1893 the company of harvesters consisted of ten men who agreed 'to work as long as the harvest lasts' for £6 2s 6d each and to be given a 'horkey' (a harvest supper) 'or not according to the behaviour and satisfaction of the Company during the harvest'.[39] In the event, the horkey was commuted into an individual extra payment of 2s 6d. The written agreement consisted of a set of 18 detailed instructions, the first of which read: 'Find a man to drive the binder to keep corners round and sharpen knives.' Another read: 'To find a man to drive the reapers to keep corners off and sharpen knives.' These two instructions are not to be found in the agreement drawn up for the harvest on the Hall Farms in 1891, suggesting that the details relating to use of machine had not then been fully grasped. The full agreement otherwise was very explicit, for example:

Agreement of Harvest on the Hall Farms 1891
The Company of Harvesters is to consist of 15 men (without the man to attend to carthorses) and the binder will be used as much as possible within reason in the wheat only.
The Company of Harvestmen will have to do as follows

1. Wheat that is cut with the binder to be set up in rows right across each field.
2. Wheat that is cut with the reapers to be tied and set up in rows right across each field.
3. All the wheat to be carted and stacked in a workmanlike manner.
4. The barley to be cut with scythes and reapers according

to the master's orders which entirely depends on the weather and the layer.
5. All the barley to be gathered by the Company.
6. All the barley to be carted, stacked and dragraked behind each cart in a workmanlike manner.
7. All the first lot of wheat rakings to be carted with the sheaves if required.
8. All the laid places in each field where the machines are at work to be mown out if required.
9. To let the reapers in every field that is required.
10. To thresh corn 1, 2 or 3 days as required.
11. The machines to be driven by men in the Company.
12. To gather, cart, stack and dragrake behind each cart all the rakings.
13. All sheaves to be set up when fallen down if required.
14. To get all stack bottoms.
15. To turn barley as many times as is thought necessary by the master.
16. To cover up stacks each night properly and put loads under cover.
17. The master finds drivers, leaders and does the horse-rakings.
18. All work to be commenced and discontinued according to the master's orders.
19. In the case of one or more of the harvest men being absent the remaining part of the company will complete his or their work and agree to payment thereof.
20. The wages of the harvestmen will be eight pounds (£8.0.0) per man to work as long as the harvest lasts.
21. It will be the option of the Master to give a horkey or not, according to the behaviour and satisfaction of the company during harvest.

Added to this agreement was a statement of the payments made in settlement. There is no record, however, of whether a 'horkey' was provided or not. A shilling 'earnest' was paid to each harvester sealing the agreement.[40]

These agreements were customary mainly in the Fen districts and East Anglia. There are few extant examples and it must be assumed that many were purely verbal as were so many other agreements. An examination of the agricultural circumstances can explain why such agreements were made in counties like Cambridgeshire, Essex and Suffolk. First, they were all large arable corn-growing districts - Cambridgeshire with 262,597 acres, Essex with 406,206 acres and Suffolk with 405,834 acres under corn crops in 1866 (compared to, say, Oxfordshire's 156,338 acres). Secondly, they had a comparatively low labour/farm ratio, e.g. Cambridge with 8 compared to Berkshire's 14 (1851). Thirdly, there was a low ordinary and high harvest wage paid, e.g. Essex 11s, but harvest pay for the month £6 with beer - compare Berkshire's 11s, harvest 16s per week (1860).[41] Furthermore, gang labour was employed in seasonal work indicating an absolute shortage of workers;[42] and Irish harvesters came to Essex and Cambridge. In this situation migrant labour was badly needed and high wages had to be paid to attract a sufficient number; so harvest agreements in a specific form controlled both methods of work and performance and provided a legal instrument enabling farmers to retain labour or prosecute for absenteeism, even for faulty work.

These circumstances did not apply to Berkshire, Buckinghamshire and Oxfordshire because of the relatively large labour pool existing in the 'open villages'.[43] Individual piece work with family help was a predominant method of taking harvest work.[44] The hired men were 'held' over the harvest period by their annual agreements,[45] and, finally, in Buckinghamshire where annual hiring agreements were much less common,[46] agreements were made individually for the harvest period - the 'month' men were thus also 'held'.[47]

Throughout this period the harvest labour situation in the south was 'sticky'. The perennial problem was to find enough labour and then to retain it. All agreements were made with this situation in mind. Harvest agreements generally solved the problem while the labourers' gain was from the satisfaction that for once they were 'bargaining the rate'.[48]

The introduction of the mowing machine, the mechanical reaper and subsequently the reaper-binder did not end the traditional methods of harvest payments but new alternative methods directly linked to their use were evolved. Labourers were paid at their ordinary day rate but extra payment was based on the acreage cut when using the machines. This direct linkage of extra pay to machines parallelled an already developing practice in the late 1860s and 1870s of relating extra pay to the time engaged in a particular task or overtime for extra hours of work as opposed to paying a general enhanced day rate, though this practice still continued. A fixed sum or extra money added to the day rate was paid for example for a day's carting, or if the harvest work was more general a fixed overtime rate was paid in addition to the day rate for the extra hours worked.[49] Examples of all these different payments can best be illustrated by referring to the payment made on a particular farm. These entries for the hay and corn harvest of 1901 on a Little Wittenham farm, Berks, are itemised in the Labour Book:

Labourers' pay for the weeks beginning July 5th

> 6 days pay 4s6d; hay cart 9d; earnings 5s3d. 6 days pay 14s; hay cart 1s6d; overtime 6d. mowing 20 acres at 2d - 3s4d; earnings 19s6d. One day hay cart 1s5d; earnings 1s5d. 6 days pay 14s; overtime 8d; earnings 16s6d. hoeing mangel 10s10d; 1 day hay cart 3s8d; overtime 8d; earnings 15s2d. 2 days and 4 afternoons 2s; hay cart 9d; earnings 2s9d [similar entries for six other workers].

For the week beginning 12 July.

> 6 days pay 14s; mowing 24 acres at 2d - 4s; extra for hay cart 1s, 1s6d, 1s6d, 6d, earnings 22s6d [a woman] 4 days 3s4d; hay cart 2d, 8d, 8d, 8d - 5s6d [similar entries for nine other workers].

For the week beginning 31 July.

> 6 days 4s6d; binding 6d; earnings 5s. on reaping 13s, 1 days pay 2s2d. loading waggons 3d; earnings 15s5d. On reaping 12s; 1 days pay 2s; loading waggon 2d; cutting round 20 acres of wheat (10s with two others) 3s4d earnings 17s6d. Women on reaping etc. earnings 15s. 6 days pay 14s cutting round 16 acres of wheat 6s; reaping etc. 6s; earnings 26s [similar entries for 14 other workers].

For the week beginning 9 Aug.

> on reaping 10s; 2 days carting 8s; loading waggon 3d; earnings 18s3d. 6 days pay 14s; road 6d (journey money); drill 6d (drilling seeds); cutting 36 acres wheat at 2d earnings 21s [similar entries for eight other workers].

Although in the fortnight commencing 31 July the reaper-binder was in use cutting 36 acres of wheat, piece work was by no means altogether eliminated. There was still the shocking to be taken at the piece-work rate of 1s 6d per acre and a limited amount of hand cutting. Though the entries refer to reaping, the hand tool in fact in use was the fagging hook, and the price paid for cutting tying and shocking some ten acres of wheat was 14s per acre in addition to 16s paid for cutting round the 36 acres in readiness for the binder. The small payment of 5s to a woman 'on reaping' is of interest, suggesting the late survival of the reaping hook on the odd occasion.[50]

Another method of payment was to take into account the use of either reaping machine or reaper-binder and establish an overall price per acre for corn cut, tied and shocked. A set of accounts for a Huntingdon estate at the turn of the century illustrates this method with the survival of the harvest 'extra' paid to the 'month' men who were responsible for all the carting and stacking. Both methods were contracts and 1s 'earnest' (6d for lads) was paid to each labourer sealing the agreement. The amount of the first was based on acreage payments for corn made ready up to the carting stage all at piece work taken jointly and shared equally by the harvest company. The 'month' men each agreed to work for an individually agreed sum for the harvest period of a month. At North and Leycourt Farms in 1897 a company of eight shared £68 8s for their harvest work using the binder for cutting all the wheat, but cutting barley, oats and peas by hand at 8s per acre, and mowing beans at 6s per acre. This payment included the 8s 'earnest' money and an extra £4 for underpricing the bean cutting. Ten 'month' men were paid various sums amounting in all to £32 1s - their month's work including the harvest work on a third farm (Hardwick) on the estate. On Lodge and Common Farms in 1903 a company of eleven shared £86 16s 3d using the binder to cut the wheat and some of the barley and oats. Most of the barley however was fagged, over half the oats, eight acres of beans and peas, while 23 acres of clover was mown. In 1897 the harvester company of eight (eleven in 1903) individually received £8 10s 9d (£7 17s 0d in 1903); the highest payments received by any of the ten (25 in 1903) 'month' men were £7 and £6 paid to two 'horse-keepers' (£8, £7 10s and £7 the three highest payments in 1903).[51]

Thus even with the introduction of machine cutting, independent individual labourers received higher payments than the regularly employed if they could find work in companies either taking the early stages of the harvest work or the whole of the harvest. So long as harvest work was divided up so that the major part of the work in the initial stages was shared by the independent labourers they held the advantage in earnings, offsetting the uncertainty of working at all during the rest of the year. Put another way, the most profitable areas of harvesting were in the early cutting, tying and shocking stages, the less profitable, the final stages of carting and stacking.

While the mower and reaper rarely eliminated piece-work cutting, the binder was to effect a drastic change in the organisation of harvest labour. Henceforth only very limited piece work was available on farms where machinery took over, for piece work was generally only available for shocking the sheaves or when machines could not cope with badly laid crops. With a decline in corn cultivation and with a growing scarcity of labour, farmers became much more concerned with maintaining a somewhat larger permanent staff who, with the aid of machines, might represent something approaching self-sufficiency in order to cope with the harvest. A new logic of placing the opportunity of extra earnings in the way of the regular farm employees developed, made possible by a decline in production. The following hiring agreement made on a Cricklade Farm in 1911 when harvest machinery had become well established illustrates the improved conditions of hired labourers.

> Oct 11 1911. Head carter engaged at 14s per week, cottage and garden free; 40s over money due Michaelmas 1912.
> 1d per acre for all drilling
> 7d per acre for mowing grass and seeds
> 9d per acre for binding corn
> 6d for every 10 qts of corn sold and delivered
> 5s for every horse sold over £50
> 2s 6d for every horse sold under £50 except screws
> 1s earnest sealing the agreement.[52]

Such an agreement can be compared with many made during the middle years of the nineteenth century such as the following, when extra earnings even at harvest time could be minimal:

Agreements on Manor Farm, Horton, Bucks for hired men.
Old Michaelmas 1850.
Cowman and half man to make himself generally useful -
to have 5s per week and £2.10s at Old Michaelmas 1851.
Shepherd to have 5s6d per week and at Michaelmas 1851 £4
as wages.
Oct 11 1851
Second team carter to have 5s6d per week and £3 at Michaelmas 1852.
Oct 11 1852
Cowman and half man to make himself at all times generally
useful to have 4s6d per week and to have £2.5s at Michaelmas
1853.
Old Michaelmas 1856
Cowman and half man to have 6s per week and £4.10s at
Michaelmas 1856 to make himself generally useful and to go
to church alternate Sundays.

In the case of these five agreements free accommodation was provided at the farmstead; the term 'half man' meant working part of the day as a general labourer. The wage itself was an all-the-year-round rate with no other cash payments.[53]

The following agreements give an indication of the amount of total earnings in cash hired men could obtain in the middle years of the century:

Agreements at Wick Farm, Radley, Berks.
1851 Oct 16. Hired at Abingdon Fair Andrew Read at 1s per
week more than day men [i.e. 9s per week] and to pay 1s
per week rent to live in the Lodge to be carter - no Michaelmas Money - no extra payments at harvest time through
piece work etc. - equivalent to £20.16s per annum net.
1853 Oct 14 Hired Joseph Squire as carter at 8s per week
and <u>£3 wages</u> at <u>Michaelmas</u> - equivalent to £23.16s per
annum net.[54]
1861 Oct 14 Hired William East - Wage equivalent to £29 per
annum net.[55]

A shilling 'earnest was paid to seal all three agreements.

With the advent of the reaper-binder, the probability of a marked decrease in the labourer's share in the initial stages of harvest created a new situation in counties like Berkshire, Buckinghamshire and Oxfordshire. It was individual piece work with family help that enabled the 'open' villagers to redress at harvest time their uncertain winter earnings. Though companies of harvesters might survive elsewhere since the use of machinery was structured into their agreements, the practice of making such harvest agreements was not common to these counties. To incorporate the use of the reaper into family methods of working was still possible but the incorporation of the binder was not. Thus more and more of the most profitable work was redistributed; the hired men were switched to working in the initial stages of harvest work and family labour in those stages of tying and shocking, which were both the most suitable and profitable work, eliminated. The gradual introduction of the mechanical mower and the mechanical reaper, with its splendidly painted sails, from the 1850s onwards was, and was accepted as, the logical process of speeding up and rendering less arduous the cutting of the corn crops in line with the replacement of the sickle or reaping hook by first the heavier fagging hook and then the scythe. First increasing production and then a decline in rural workers ensured that the labour force was still extended so that availability of work was not curtailed. But the mechanical binder presented a threat to a pattern of family participation which had existed from time immemorial. And it was this threat, not the threat of strangers or Irishmen to their livelihood, that brought the villagers out into the streets of East Hendred at the turn of the century when this machine was first introduced on the Allin Estate. The farm workers held torchlight protests and even set a thatched building ablaze.[56]

The shift in the economic advantage of harvest work away from the independent labourers is reflected in the increasing proportion of hired men. A continuing independence, precarious though it might be, was to become increasingly difficult. More regular employment, if available, was offset by the restrictions among others of inhabiting a 'tied' cottage. As a Britwell Salome villager put it at the turn of the century: 'You didn't dare say a word as big as a clover seed or you might lose your job. There was always someone else waiting for it.'[57] This was as much an expression of a new social grievance as an old economic one.

The situation that emerges can be summarised thus: in the period of intensive hand-tool usage travelling bands, whether Irish or English, and specialist labour with its host of followers took the major portion of the lucrative piece work. With the advent of machinery this situation was slowly transformed to the advantage of the hired men. The introduction of machine mowing and reaping between 1850 and 1870, introduced in a period of increasing production, saved time but did not replace labour; it released labour for other operations such as hay making, carting etc. From 1870 onwards, machine mowing and reaping brought less piece-work cutting and a decline in migrant and immigrant labour - travelling harvesters and disadvantaged day labourers - with the extra payments going to hired men who worked the machines. But still ample work was available at hay making and carting for day labourers, women and followers - tying, shocking and carting in the corn harvest. There was still very considerable use of hand-cutting tools but it was on the decline. From 1880 onwards, the introduction of the machine binders, its slow adoption increasing by 1900, brought elimination of both cutting and tying. The greatest effect was on direct or indirect employment for women and children, the decline in piece-work earnings of day labourers and piece-work shocking shared with the hired men. There was an increase in the regular labour force and a decline in the need for specialist casual labour, with extra payments to hired men. It was the destruction of family work and the cause of decline in the numbers of agricultural workers who could survive as independent day labourers.

NOTES

1 F. G. Heath, 'The English Peasantry' (1874), p. 187. 'Nothing short of combination would effect any improvement in the deplorable conditon of the peasantry' (quoting Cannon Girdlestone at the British Associaton meeting, Norwich, 1868).
2 E. L. Jones, 'Seasons and Prices (1964), p. 61. 'The carrying of the harvest... remained as the central operation of the farmers' year. In late summer hands were therefore at a premium...'
3 Oxford RO, Dillon Mss. DIL 1/q/7b.
4 Jones, 'Seasons and Prices', p. 141.
5 Ibid., p. 143.
6 Oxford RO, DIL 1/q/7i.
7 Ibid., DIL 1/q/7n.
8 Ibid., DIL 1/q/7r.
9 Ibid., DIL 1/q/9c.
10 A. Wilson Fox, Agricultural Wages in England and Wales During the Last Fifty Years, JRSS, vol. LXVI, pt. II (1903), pp. 273ff.
11 'Morning Chronicle Supplement', 21 Dec. 1849, Rural Districts, Letter II. Estimate based on four counties including Oxfordshire.
12 A. L. Bowley, The Statistics of Wages in the United Kingdom During the Last Hundred Years, JRSS, vol. LXII, pt. III (1899), pp. 555ff.
13 The longer hours of late spring, summer and early autumn days from 'dawn to dusk' were formally recognised by usually paying an extra shilling per week during the second half of the annual engagement.
14 Reading Farm Records, BER 13/5/2, Wick Farm, Radley.
15 Ibid., OXF 6/1/3, Manor Farm, Aston, 12 Oct. 1883.
16 See ibid., BUC 7/1/1, 1860 harvest rates: ordinary rate 10s, enhanced rate 15s; £1 13s double rate; 3s a day (time and a half). 1861 harvest: day rate 2s, enhanced rate 3s 6d (at Hill Farm, Stokenchurch, Bucks).
17 Oxford RO, DIL 1/q/9c.
18 Ibid., DIL 1/q/9b.
19 Reading Farm Records, OXF 2/2/4, ff.22-177.
20 PP 1868-9 (4202) XIII, RC Employment... in Agriculture (1867), p. 150 'as they are generally in charge of horses, cattle or sheep, their opportunities... are... confined to about a fortnight's reaping or fagging' (Oxfordshire).

21 Ibid., p. 151 'the carter is usually allowed by piecework in harvest to earn 25s extra' (Berkshire).
22 Reading Farm Records, BER 13/5/5, Labour Book.
23 See Warwickshire Labourers' demands preceding Wellesbourne strike 11 March 1872: '2s8d per day...; hours from 6 to 5; and to close at three on Saturday; and 4d per hour overtime'. Heath, 'The English Peasantry', p. 197; cf. G. E. Fussell, 'From Tolpuddle to T.U.C.' (Slough, 1948), p. 68; but note a warning article in 'Labourers' Union Chronicle', 7 June 1873, p. 6, 'a mistake to ask for increase in wages and decrease in hours at the same time'.
24 PP 1881 (C 2778-II) XVI, RC Agricultural Depression, p. 324 (vi).
25 See Chapter 9.
26 Reading Farm Records, BUC 7/1/1, £21, 19 Sept. 1850.
27 Oxford RO, DIL 1/q/9c. See also Reading Farm Records, BER 28/3/1, Chieveley Farm 1869, harvest home money: 18 men, 2s 6d each, 10 women and 3 lads, 1s 6d each; Bradley Hill Farm, 18 Oct. 1872, harvest home money paid, £2; 23 Oct. 1875, £1 7s 6d (Account Book - same estate).
28 Oxford RO, OXF 14/3/5-10. Reading Farm Records, BUC 7/1/1, Hill Farm, Stokenchurch, Bucks: 28 Aug. 1849 provisions for men, 3s (£1); 24 Aug. 1850 provisions for men carting, 5s2d (£19); 19 Aug. 1851 men's beer, 4s (£41); 30 Aug, beer/extra hands, 6s4d (£42). Beer supplied free all the time, average 6d to 8d per day.
29 Oxford RO, 6/1/3, Labour Book entries for 28 Sept. 1883.
30 See ibid. for the otherwise poor ordinary and harvest rates paid to women.
31 Individual piece work was competitive for it exhausted work opportunities.
32 C. Henry Warren, 'The Happy English Countryman' (1955), p. 68.
33 Arthur Randell, 'Sixty Years a Fenman' (1966), p. 23.
34 G. E. Evans, 'Ask the Fellows who Cut the Hay' (1956), p. 85.
35 A. F. J. Brown, 'English History from Essex Sources, 1750-1900' (Chelmsford, 1952), p. 40.
36 See Appendix A, p. 195.
37 PP 1868-9 (4202-II) Bi, p. 581 (41).
38 Ibid., p. 581 (39).
39 Reading Farm Records, CAM 1/1/2.
40 Ibid., CAM 1/1/1.
41 'Ag. Gaz.', 30 Apr. 1860, p. 392.
42 Agricultural Gangs, 'Quarterly Review', vol. 123 (1867), p. 174, 'in Lincolnshire, Huntingdonshire, Cambridgeshire, Norfolk, Suffolk, Nottinghamshire... limited degree in the counties of Bedford, Rutland and Northampton'.
43 Only ten counties had a higher proportion in rural occupations than Oxfordshire (27.1%),
44 PP 1868-9 (4202) XIII, p. 82 (47), 'every available man, woman and child... is at work... in haytime and harvest...

it is the custom for families to work together... taking the reaping and binding by the acre'.

45 But see PP 1868-9 (4202) XIII Bj L594 (97). 'In the Woodstock Union farm servants hired by the year are not infrequently taken before the magistrates for breach of contract.'

46 PP 1867-8 (4068-I) XVII, RC Employment, f.p. 523, 'very little yearly hiring in Aylesbury Union', and p. 526, 'Only single men hired for a year' (Upper Winchendon).

47 For rates for month men, see 'Ag. Gaz.', 30 Apr. 1860, p. 392: 20s per week with beer (Claydon), 24s (Bladwell), 26s (Woolston) for 1860.

48 Note piece work itself was a contract of work, but see Chapter 8 for the breaking of agreements.

49 Reading Farm Records, BER 10/2/1: 5 Sept. 1879, 8 entries for piece-work 'fagging', overtime 6d; 26 Sept. 8 entries for piece-work 'fagging', 5 entries 'day at cart' 5s 8d and 4 others at 4s; 4 entries double harvest pay for the week, 9s, £1 18s, 18s, 12s, Little Wittenham, Berks.

50 Ibid., BER 10/2/2.

51 Ibid., HUN 2/1/1, W. B. Fowler, Great Gransden - Account Book 9 Aug. 1897 - 21 Aug. 1916.

52 Ibid., OXF 6/1/3; note 'screw', a vicious, unsound or worn-out horse, term for any emaciated farm animal.

53 Ibid., BUC 1/5/1, Labour Book.

54 Ibid., BER 13/5/3 (1853).

55 Ibid., BER 13/5/3 (1861).

56 Edwin Manley, 'A Descriptive Account of East Hendred' (Oxford, 1969).

57 G. E. Moreau, 'The Departed Village' (1968), p. 66.

8

SOCIAL DISCONTENT AND DISCORD

DISCONTENT

> A spirit of discontent has sprung up among the labourers respecting their cottage accommodation and small pay. They demand higher wages and shorter hours of work. Guided by men more astute than themselves they determined to make the harvest the testing point of the contest; they arranged themselves that 'if you, the farmer, do not pay us higher wages we shall neither mow nor reap'. The question has lately come to a crisis in various counties. Labourers in many instances have struck work. Agriculturists have been obliged to advance wages in order to get work done, but discord still prevails between masters and men.
>
> 'Ag. Gaz.', 24 Aug. 1872, p. 1144.

The Repeal of the Corn Laws did not bring disaster to either landlord or tenant farmers, but during the 'Golden Age' of the middle years of the century the economic gap between the increasing prosperity enjoyed by farmers and the relatively static condition of poverty of the agricultural labourers widened: 'the gold had gone to the landowners, the silver to the farmers and only the coppers to the labourers'.[1]

The degree of this change was captured by an opinion current among older villagers based on their observation of succeeding generations of farmers' children: 'It used to be Mom and Dad and Porridge, and then 'twas Father and Mother and Broth, but now 'tis Pa and Ma and Soup.'[2] Farmers indeed thrived. The occupant of Idstone Farm made considerable profits during the Crimean War which enabled three sons to take up large farms during the 1860s.[3] A Newbury butcher's dealer thrived and by 1890 owned several farms and estates including Harefield Farm, Hendred, Shelsmore Farm, Donington, Ashfield Farm, Newbury and The Rookery, Henwick - at one time over 1,600 acres were in his possession.[4] Even though average wheat prices had fallen from 17s 6d per cwt (1855), the price was held at 15s and 14s 11d in 1867 and 1868, while barley prices in five years during the 1860s were above the 9s 9d per cwt of 1855. Increasing yields and acreages in corn more than compensated for any fall in price.[5]

Consumer prices on the other hand were almost at their peak, cutting into the labourers' low income, and when the Education

Act of 1870 was passed, labourers saw a double threat to their already unbalanced budgets in the payment of fees and a restriction on their children's earning power.[6]

If little corporate action had taken place between the destructive movement of 'Swing' and the constructive movement of 'Arch' there had been individual action. Those labourers with an abiding discontent migrated or emigrated but in the process they lessened the chances of a direct confrontation with farmers; the initiative to challenge them was often lacking in those remaining who were too old, too fearful, or too tied to leave familiar work and familiar faces for an unknown, uncertain life elsewhere. A contemporary Oxfordshire saying was:

> When you've got one you may run
> When you've got two you may go
> But when you've got three you must stop where you be.[7]

Moreover, it needed far more than haphazard individual migration to reduce the surplus pool of labour which kept wages down.

But when organised efforts were made to encourage migration, farmers made clear their antagonism. In a speech in 1867 made at Tiverton in Devon, Canon Girdlestone the pre-eminent protagonist of migration gave details of his success in turning the law of supply and demand, for once, to the advantage of the labourers in his own parish of Halberton.

> By my instrumentality no less than 38 labourers from this neighbourhood - about half of whom are married with families, and half single men - have been removed to various places - Ireland, Lancashire, Yorkshire, Essex, Kent etc., where the married men all get house and garden rent-free, with sometimes fuel added, with wages varying from 12s a week, the lowest, to £1 a week; and the single men mostly board, lodgings, washing and mending, and 6s a week. This is still going on... the vent is made, the farmers' dictum, which they thought I was too ignorant to know about, and too wanting in power to make practical, is fulfilled. Wages are finding their own level...

For his efforts to temper the discontent resulting from low wages the Canon received a merciless pelting from the farmers and the newspapers which espoused their cause.

In the same year, however, labourers at Gawcott in Buckinghamshire received more than obloquy for their own efforts to remedy low wages. They struck for better wages in March demanding 12s a week and 1s for Sunday work, having worked all the winter for 9s and 10s a week. They were refused and promptly paid off.[8]

If farmers were adamant in their opposition to migration and in their refusal to consider wage demands, there was a further major grievance in the harsh manner in which the laws relating to work agreements operated. For the workers, harvest represented work opportunities and a chance to exercise some freedom to reject unfavourable conditions or unfair rates of pay in a wage economy. Farmers accepted the advantages of a free wage economy at all times except at harvest; their actions then reflected a feudal attitude demanding from workers a dedicated view of the harvest however small the labourers' share in the fruits of harvest were to be. What a community of interest might have effected in the past the farmers had to resort to the Master and Servant Act to achieve in the changed circumstances of nineteenth-century agriculture. Until 1875 it was not only a civil but could be a criminal offence leading to upwards of three months imprisonment for 'aggravated breach of contract'. Although this act concerned all forms of labour the particular case with which aggravation could be claimed as occurring in agricultural employment until this date rests on the direct link of the various statutes dealing with hiring labour - the Elizabethan statute commonly called the Statute of Labourers (1563 5 Eliz. I c.4 Artificers and Apprentices Act). This act stipulated that: 'All persons not engaged in handicrafts or serving another, and without independent means, were compelled between the ages of twelve and sixty years to work in husbandry, for whosoever requested them.'[9] And more relevantly it enacted that:

> [I]n the time of harvest, for the avoiding of the loss of any corn or hay, justices or constables or other head officers may require artificers and persons meet for labour, to serve by the day in mowing, reaping, shearing, getting or turning of corn, grain, or hay... upon refusal to put him in the stocks for two days and one night... women between the ages of twelve and forty, being unmarried might be compelled to serve in such employment as the justices might direct, under pain of imprisonment.[10]

The Statute of the 20 Geo. II c. 19 (1747) gave 'Summary jurisdiction to justices of the peace in the matter of disputes between master and servants' hired for one year or longer (extended by 31 Geo. II c. 11 53 1757 to those hired for less than a year), but compulsory recruitment of labour was abolished. If a complaint was upheld, punishment could be either some abatement of wages due or committal to the house of correction 'there to be corrected', by whipping and 'to be held to hand labour for a reasonable time not exceeding a month', or dismissal. Dismissal as a punishment under this act rather than leaving service as a reason for prosecution suggests that 'misdemeanor, miscarriage or ill-behaviour' were the matter of the moment in the eighteenth century. But this was not so in the mineteenth century for the next general statute (4 Geo. IV c. 34) in 1824 dealt with breaches of contract 'in not entering into the service at all, or in quitting it before the term agreed on had expired', as well as 'any misdemeanor or misconduct', and provided justices of the peace with powers to impose terms of imprisonment not exceeding three months. It was under this act that farmers could act punitively and pressed for imprisonment whenever hired men broke their agreements or whenever a dispute arose at harvest time.[11] The Act of 30 931 Vic. c. 141, entitled 'The Master and the Servant Act 1867', theoretically improved the worker's position for imprisonment could no longer be imposed for a simple breach of agreement. The fact however that farmers could plead that any withdrawal of labour put crops at risk meant that offending labourers themselves were at risk under Section 14 of the act which now made a distinction between a simple breach and one of 'aggravated character'. Charged under Section 14 a labourer could still be sent to prison for upwards of three months.

Table 8.1: Labour Law Convictions

(Return of the number of persons summarily convicted and committed to prison for breach of contract in neglecting work or leaving service during each of the years 1854 and 1855 (1824, 4 Geo. IV c. 34)

	1854	1855
Berkshire	73	47
Buckinghamshire	22	11
Oxfordshire	26	7
England	2,427	1,541

(Return of all convictions under the Criminal Law Amendment Act, 1871 (34 & 35 Vic. c. 32) between 29 June 1871 and 8 May 1873)

Buckinghamshire	At Newport Pagnell, 7 May 1872 - obstructing a labourer with a view to coerce, 7 days imprisonment.

Sources: A & P 1856 (441) L, Law: Police, pp. 633-4; A & P 1873 (385) LIV, Law and Crime, Criminal Law Amendment Act.

NB: Such offences are to be distinguished from those leading to prosecution after 1875 under the Conspiracy and Protection of Property Act (38 & 39 Vic. c. 86), which, though rare, carried over the threat of imprisonment previously contained within the Master and Servant Act (Sec. 14) - see Table 8.2, p. 127 - the Act of 1824 (up to 3 months), and the less harsh act (31 Geo. II c. 11) of 1757 (up to 1 month). The relevant clause of this act, still on the statute book, was Sec. 5;

> In some cases the breach of contract of service will amount to a criminal act... a person who willfully and maliciously breaks a contract of service or of hiring... to endanger human life... or to expose valuable property to destruction... liable to a fine of £20 or to imprisonment for three months with hard labour ('Encyclopaedia of the Laws of England', vol. VIII (1898) p. 260, see also ibid., pp. 235-65).

Frederic Harrison, a radical lawyer, very quickly took issue over the manner in which this section was being used, citing the case of an Essex labourer who

> earning 7s a week for 14 hours' labour left his valuable situation for work in another county; on return six months later he was sent to prison. Another man on asking for a rise in wages was told that he could leave if he did not like his wages. On taking his employer at his word the labourer was summoned to appear at the Petty Sessions - the employers regarded this too scrupulous literalism in carrying out orders as 'aggravated misconduct'.[12]

Between 1824 and 1875, therefore, labourers risked a penalty of upwards of three months imprisonment if agreements were broken; breaches of agreements, however, on the part of the employer ranked as civil offences - fines or restitution for unpaid labour being the only penalties. A further statute in 1875 removed this inequity, but if under this, the Employers and Workmen Act (38 & 39 Vic. c. 90), labourers' breaches of agreement no longer carried criminal liability, the threat of three months imprisonment still remained within the scope of the Conspiracy and Protection of Property Act (38 & 39 Vic. c. 86) enacted the same year.

Immediately prior to the passing of this act in July 1875 a deputation from the National Agricultural Labourers' Union saw the Home Secretary because of the inevitable consequences of a clause in the act which put labourers in the country in a worse position than the labourers in the town:

> What they wanted was words in the fifth clause to show a breach of contract should be treated civilly... simple disputes as to hours of labour, wages, or anything of that nature should be treated civilly and not criminally...

The Home Secretary was not to be moved: 'Supposing in weather like this a man left haymaking... the probable consequences would be the loss of the crop.'[13] This then was the legal background regarding the engagement of all labour. Although the hired man had some security he was always in constant fear of dismissal. For the average farm worker in rural England there was little economic freedom. It is true, restrictions on the activity of farmers were contained in their tenancy agreements and conventions certainly restricted their social behaviour in the tightly knit hierarchical structure of Victorian rural England, but the full weight of social, economic and legal strictures fell on the agricultural workers.

A word out of place, let alone unacceptable behaviour, might quickly bring economic disaster, for dismissal from employment meant not only the loss of a job but the loss of the family dwelling.[1] A labourer's comment made to the 1893 commissioner encapsulates the feelings of perpetual constraint created by farmers in their hired men:

> some of us dare not call our souls our own, if we ask for a contract for piece-work we get the sack at once, we do not get a week's nor a day's notice, it is a minute's notice which we get, and the farmers do not care what becomes of us.

Another labourer from the same area in Buckinghamshire informed them: '[I]f you want a well-bred poacher and a discontented devil go to Shabbington... Farmers don't find work enough for us.'[15] Nevertheless, at harvest time, despite the farmer's acute need for labour and the labourer's equally acute need for work, the harmony of interest was often broken. The hired men might not seem to present any problem since if they broke their agreements they stood to lose their anxiously anticipated Michaelmas money. There was no such economic incentive for the day labourers or casual

Table 8.2: Labour Law Convictions

(Return of the number of convictions in each year under Section 14 of the Master and Servant Act, 1867)

	1868	1869	1870	1871	1872
Berkshire					
Lambourne	1 (3 mths)				
Buckinghamshire					
Linslade				1 (7 dys)	
Oxfordshire					
Wootton North	1	4			
	(longest 7 dys, shortest 4 dys)				
Ploughley					1 (10 dys)
Watlington	1	1			
	(longest 7 dys, shortest 4 dys)				
Banbury & Bloxham					1 (1 mth)
Henley	2				1
	(longest 2 mths, shortest 28 dys)				

(Return between 1 Jan. 1873 and 1 Apr. 1874)

		Term of imprisonment	
		Longest	Shortest
Berkshire			
Ilsley	1	11 mths	
Reading	3	14 dys	14 dys
Forest Division	1	1 cal. mth.	
Buckinghamshire			
Newport Pagnell	3	7 dys	7 dys

Sources: A & P 1873 (386) LIV, Law and Crime, pp. 227-8; A & P 1874 (360) LIV, Law and Crime, p. 395.

Table 8.3: Offences Relating to Servants, Apprentices or Masters in England and Wales, 1857-1900[1]

Year	Total	Berks	Bucks	Oxon
1857	9,687			
1858	8,301	119 (5)	53 (0)	102 (0)
1859	9,891	126 (6)	73 (3)	77 (4)
1860	11,938	152 (8)	110 (1)	90 (0)
1861	10,393	145 (3)	68 (1)	88 (0)
1862	7,637	142 (13)	58 (0)	78 (0)
1863	8,504	98 (9)	47 (0)	66 (2)
1864	10,246	145 (8)	59 (3)	73 (1)
1865	10,412	101 (14)	74 (4)	58 (1)
1866	12,345	131	73 (2)	69 (1)
1867	9,953	146 (11)	68 (2)	94 (1)
1868	8,204	91 (7)	67 (5)	78 (0)
1869	7,385	90 (2)	24 (1)	62 (3)
1870	8,670	74 (5)	32 (0)	47 (0)
1871	10,810	62 (1)	22 (5)	40 (0)
1872	17,082	110 (16)	44 (1)	91 (1)
1873	16,230	122 (3)	35 (0)	125 (1)
1874	13,544	113 (7)	38 (2)	82 (3)
1875	14,353	122 (4)	45 (3)	73 (1)
1876	9,382	88 (3)	54 (0)	37 (3)
1877				
1878	5,607	71 (8)	21 (0)	23 (0)
1879	5,508	79 (0)	19 (0)	35 (9)
1880	6,367	60 (8)	1 (1)	39 (8)
1881	6,765	30 (6)	1 (1)	18 (6)
1882	8,083	66 (5)	2 (2)	18 (6)
1883	7,321	50 (8)	5 (4)	36 (9)
1884	7,766	69 (11)		26 (4)
1885	6,072	33 (2)	1 (1)	17 (4)
1886	4,892	43 (8)	1 (1)	16 (3)
1887	5,368	38 (8)	0	20 (2)
1888	5,929	19	0	20 (7)
1889	7,776	21 (1)	0	18 (6)
1890	7,323	30 (2)	0	9 (3)
1891	8,495	26 (1)	0	23 (3)
1892	5,630	48 (9)	0 (2)	21 (1)
1893	824	2	1	0
1894	389	0	2	1
1895	235	1	0	0
1896	173	1	0	0
1897	680		0	0
1898	192	6	1	0
1899	269	13	0	0
1900	169	11	0	0
1901	53	0	0	0

Note [1] : (to Table 8.3)
Undifferentiated until 1893; specifically breach of contract from 1894 (Urban offences in parentheses but included in annual total). Related acts as follows; 1824, 4 Geo. IV c.34; 1867, 30 & 31 Vic. c.141 (The Master and Servant Act); 1875, 38 & 39 Vic. c.90 (Employers and Workmen Act); 1876, 38 & 39 Vic. c.86 (Conspiracy and Protection of Property Act).
Sources: Judicial and Miscellaneous Statistics.

workers. If they were free to starve in winter for lack of work, they were free to some extent to reject work and risk the consequences when there were various opportunities for alternative employment with other farmers. Having placed reliance on their labour, however, farmers could be gravely inconvenienced and face economic loss if harvest work was delayed or disrupted by contracts being broken and labourers leaving work uncompleted. For various reasons and on numerous occasions farm workers did behave in this way.

Farmers could not afford, for both social and economic reasons, to ignore such a show of independence, for it could affect not only their pockets but the docile attitude of other workers. They were not slow therefore to prosecute such workers as much to emphasise their authority as to show that authority spurned brought inevitable punishment. The Judicial Returns, only available from 1857, indicate the significant numbers brought to court especially until 1875.[16]

Whether through wider contact with other workers from the more independent 'open' villages or from an envy of their less dependant existence, hired men seem prepared to rebel against their conditions of employment despite the economic incentive of Michaelmas money not to break their hiring agreements, and perhaps the more disastrous consequence of prosecution. From a very small random selection of 14 cases of leaving employment, nine related to hired men and only five to day men, though more than five day men were involved. It is true that the breaking of agreements by hired men was, from the farmer's point of view, much more serious in its consequences and prosecution of other workers might go by default, so that a much fuller analysis of many more cases would have to be made for a firm conclusion.

It seems, however, that a different attitude to working conditions existed at harvest time, and the traditional subservience among hired labourers then if at no other time was less overt than is generally supposed. An indication of this difference in attitude can be shown by looking at some of the prosecutions. In September 1875 James Shorter, a hired labourer, appeared at the Bullingdon Divisional Petty Sessions in Oxford charged with not fulfilling his contract of service. On the 3 August the defendant had been ordered to cut a road through some oats for the passage of a machine but he did not do so, and damage costing £2 was caused. Shorter's objection to cutting the road was that ' he would not work by the day when others were working by the piece'; as no contract existed and no assent had been given to the rules (drawn up by the bailiff the previous March, when he came) the case was dismissed.[17] In the previous August John Palmer was summoned for leaving the employ of his master Mr Thomas Latham, farmer of Wittenham. The defendant was engaged by the complainant the previous October at 10s a week except during harvest when he was to receive 2s a week extra and a bonus of 20s at Michaelmas. He was fined 5s and 15s costs and ordered to return to his employer's service.[18]

A day labourer, Thomas Alder of Denchworth near Wantage, in July 1860 was summoned by Mr Walker of the same place for refusing his lawful commands, 'he being hired to serve through the hay-making and harvest time'. Alder denied that any hiring had taken place, but admitted refusing to comply with his master's orders; he was fined 5s and the contract was ordered to be dissolved.[19] A sign of the times may perhaps be distinguished in the case brought against four hired men at Wantage Petty Sessions on 8 September 1875, for one of the grievances agitated against by various agricultural unions was shorter working hours and the definition of the working day as so many hours work - a loose definition of from 'dawn to dusk' being no longer acceptable.[20] John Brown of East Hockings had summoned these four men 'for refusing to do harvest work after 7 p.m. on the 27th [August] and on subsequent days', significantly, the case was dismissed.[21] Specific indication of union activity in 1875 if not of influence is reflected in a case at Salisbury. Six labourers were sentenced to 14 days imprisonment at Salisbury in the first week of August for having left their work during hay harvest; the local branch of the union subsequently organised a demonstration against a magistry consisting of farmers and landowners.[22] Thomas Blakwell of North Cerney was charged with absconding from the service of Mr Robert Garns, farmer of Aldsworth, on the 12 June 1859; he was convicted and committed to the House of Correction at Northleach for one month with hard labour for this offence.[23] John Pinker of Gawcott charged by George Attenbrow of Stratton Audley with having left his service, 'he having been hired on the 11th of August to reap a quantity of wheat at the parish of Stratton Audley', which he neglected to do, did not appear. 'Service of summonse was proved and he was committed for 21 days to hard labour in September 1847.'[24]

In the first week of September 1875 William Chapman and Simon Harris, labourers of Aylesbury, were 'summoned for neglecting to fulfill a contract entered into with Mr. T. J. Hind to cut some beans at Burton having undertaken to cut 4½ acres of beans at 11s 6d an acre'. They were convicted.[25] A week later at Camden Special Sessions Wm. Smith, a farmer from Willicote, charged four men named Wm. Sharlot, Wm. Hadland, Chas. Winstone and Geo. Lines with refusing to fulfil a contract to cut wheat at Clifford on the 24 August. They had to pay 3s 6d each and 2s 6d each compensation.[26]

The working conditions of Lovegrove, a servant in husbandry

working for Mr Lot Ginger of Buckland, must have been grim. Hired on the 24 October 1855 to Michaelmas at 6s a week with £5 at Michaelmas he was charged with leaving his master's employment in August 1856, thus losing his claim to his Michaelmas money. The implications of his position and a comparison of his wages with rates obtainable in the surrounding district is best illustrated by a case heard a week previously at Aylesbury Magistrate Chambers. Here John Weedon was charged by John Deverell of the Grange Farm, Weston Turbille, with leaving his service. This lad had been hired at 4s 6d a week in the winter and 5s 6d a week for the summer, and was to receive £1 at haytime and £2 for the harvest 'but wanted some of his wages... not due'. Rev. C. E. Gray, one of the magistrates, said the defendant had no right to the wages for they were the security that he would remain 'at the time he was most wanted', and accordingly fined the lad 10s 6d.[27] A more unusual case because it occurred in winter was heard at the Aylesbury Petty Sessions on 28 December 1839. Possibly with more reason than the lad John Weedon, another lad engaged from Michaelmas to Michaelmas at 4s per week with the prospect of an extra £2 at the end of his year's service absconded on account of the carter: 'I can do nothing right for him, he's always swearing at me' said the crying boy in court. Mr Piggott the farmer had no sympathy for him, 'make an example of him, send him to prison', he demanded.[28] George Griffin, a servant in the employ of Mr Mace, farmer of Sherborne, near North Leach, did get sent to prison in August 1859. He was 'charged with refusing to assist in loading some wheat in a field on the 6th [August]' and received one month's hard labour.[29] George Austin on 21 August 1875, however, charged with unlawfully absenting himself from service had his case dismissed; as there had been no written contract, the agreement was held to be invalid.[30] A similar verdict was given for the same reason in the case of Frederick Harris, labourer of Sugworth, at the Abingdon County Bench in August 1890 but damages of £1 claimed by Thomas J. Lord, the farmer, were awarded. Harris, working for a week and a half, had left 'a field of oats partly cut' and a machine had to be hired to do the work.[31]

In these two cases and others involving doubt as to the existence of any written contract the magistrates were either being lenient, choosing to ignore the letter of the law or ignorant of the details of the law, for no contract's validity rested solely on written agreement. A binding agreement would have existed whether 'the contract be expressed oral or in writing, and be a contract of service personally to execute any work or labour'.[32] It might be said that in deciding the cases on a technicality though a wrong one, a direct verdict against the farmers was avoided. It might indeed be said that actual signed agreements were not normal practice. This was in fact the opinion reached by the commissioners in 1893.

> [T]he Contracts are not very often in writing but they generally contain an agreement to pay Michaelmas Money

> which is a means of holding labour. If these hirings are broken the M.M. is in some cases forfeited, but where there is a written agreement the amount to be paid is often settled by the magistrates.[33]

If wage agreements and harvest contracts were broken by agricultural workers, farmers were no less guilty in attempting to avoid paying for harvest work. And contrary to general opinion,[34] labourers in their turn charged farmers before the magistrates if payment was withheld. A good illustration of farmer suing labourer and then labourer suing farmer occurred in 1859. At Drill on Friday, 9 September the case of Mr A. Harvey, farmer, of Chilton v John Bass, labourer of Drill was heard at the Magistrates Meeting.

> A verbal agreement was made between Mr. Harvey and Bass (jointly with others) to cut Mr. Harvey's crops <u>rough and smooth</u> at 12s per acre. Bass had worked four days and cut his part of the harvest work, and he had commenced on the lighter work when Mr. Harvey had put fresh hands into it; this displeased Bass and he left. Mr. Harvey said he was compelled to give at the rate of 15s per acre for the part Bass would have cut had he not left. As no positive agreement was shown to have existed between the two parties, the case was dismissed. Mr. Harvey reluctantly paid expenses.[35]

A fortnight later on Saturday, 24 September Richard Harvey and John Bass again appeared together, this time at the Thame County Court, this time with Richard Harvey as defendant. Again the proceedings are fully reported:

> John Bass v Richard Harvey, farmer. The plaintiff is a labourer residing at Chilton and sued the defendant, a farmer living at the same place for the sum of £1. 0s.6d due to him for cutting corn. In defence it was stated that Bass, after he had done work to the amount claimed, left his employment and got work on another farm to the inconvenience of Mr. Harvey. Plaintiff however proved that he was not engaged to do any given amount of work or to stay until the whole of the reaping was done, and further that the reason he left was because Harvey put him (the plaintiff) on the <u>heaviest</u> work, with the promise on completing the work he was on, he should have some of a light sort, instead of which defendant's own men, who worked by the day, were put on the light work.

In summing up, the magistrates decided there was no breach of contract, 'no proof... that one had been made,... [made] an order for the amount and 2s6d for loss of time' adding they 'would advise defendant not to employ men again without coming to a proper understanding'.[36] If the occasions when labourers sued farmers were unusual, they were certainly not rare. At Chipping Norton County Court on 3 August 1865 a Chipping Norton man claimed £1 13s for labour in getting in hay. The farmer contested the claim on the ground that 'the hay was rendered unmarketable through plaintiffs negligence'. Judgement for 22s without costs was made in favour of the plaintiff.[37] At Faringdon Petty Sessions in October 1890 a carter summoned a farmer for wrongful dismissal. The plaintiff said that the defendant hired him on 1 April, the verbal agreement being that he was to continue in the defendant's employ till Michaelmas at which time he was to receive £1 over money. The defendant gave him 1s earnest money. 'On September 20 he was discharged. He therefore claimed the £1 over money and two weeks' wages at 12s.' The defence was that the plaintiff had been engaged 'by the week'. The magistrates found for the plaintiff but he was only entitled to the £1.[38]

Many and various indeed were the excuses made by farmers to justify their refusing to pay their men. Joseph Hazell, on the point of completing his annual hiring, was 'dismissed because he refused to cut chaff'. Michaelmas money of £3 was claimed by him at the Amersham Petty Sessions on 28 October 1889. A compromise settlement of 30s was made.[39] Two labourers were refused payment for mowing six acres of beans at 6s an acre in the harvest of 1861 because 'they left off work at three o'clock on a Saturday afternoon to go to market, intending to finish the work on the following Monday'. But the farmer would not allow them to finish the job. The Brackley County Court judge ruled it 'a very frivolous excuse'. The labourers received 30s, the whole amount due, 'and 3s each for witnesses'.[40] Two Barton-on-the-Heath labourers were refused a payment of £5 11s 0d for mowing grass because the farmer claimed that they had disobeyed his orders and 'had drunk his cider extravagantly'. The judge at Chipping Norton County Court sitting on Tuesday, 5 August 1862 gave judgement against the defendant for £5 1s 0d and costs.[41] In August 1889 a Wardington labourer claimed £1 15s 9½d from a local farmer for work done, but there was a counter-claim. The labourer was allowed 15s 9½d with costs. The counter-claim was then investigated. W. Judge, the farmer, claimed 11s 8d for seven gallons of beer. It was ruled that he could not recover for the beer as he simply gave it 'for his own convenience' and 'to whip the men up'; a sum of 4s also claimed for 'rakes etc.' could not be recovered.[42]

Three strands of action can therefore be distinguished in the years prior to the sudden emergence of unionism in the winter of 1872. Discontent had induced individual migration and abortive strike action, but also a continuing individual challenge to the farmers' legal rights over their labour force. Arch's Union created an organisation and a degree of solidarity to ensure some degree of organised migration. Success in these areas of conflict for strike action, sustained into the months of harvest by a rapidly growing body of labourers, was not only legal but a much greater disruptive force than individual decisions to migrate or the illegal breach of an agreement by isolated individuals. That it did continue into the harvest months, however, was as much due to the weight of farmers' opposition to consider the justice of any change in the labourers' social or economic conditions.

The farmers very early on sensed the new mood in the countryside. Before February was out some masters were offering a rise of 2s to prevent their men from joining the Labourers' Union,[43] but a strike was already in progress on the Ratley Estate the day of the Wellesbourne meeting - a number of carters, shepherds and labourers leaving their work without notice. Lord Jersey, however, intervened giving notice to those men who occupied his cottages unless they returned to work; the masters also offered to pay them for the time they were out. Originally the dispute was over beer considered not fit to drink on one of the farms where the men were threshing. The men agreed to ask for better beer and more money, the master offered only better beer which was refused, and so the men were told they might go: 'then the strike commenced'. Under notice to quit their cottages but with an offer of 1s extra 'some... returned and some... not... some... gave a fortnight's notice to leave again... and some left again'.[44]

The agitation gained momentum in March. Wellesbourne strikers demonstrated at Middle Tysoe on Tuesday 19 March forming a branch society for the surrounding district including Brailes, Shennington and parishes on the Oxfordshire border. 'Already about 1,500 men had joined the union', announced Arch, 'the farmers were unwise in putting on "a screw", they would be in difficulties in harvest time for want of labour.'[45] The worried masters of Stoneleigh met their labourers at the end of March and agreed to raise wages to 15s a week. At the same time, at a union meeting at Kenilworth, emphasis was put on old age to gain recruitment: 'It was in many districts the common lot of the agricultural class to look forward to the Workhouse as their doom in old age: Unionism would prevent this.'[46]

All this time leading articles, special editions and a spate of letters appeared in the local newspapers,[47] and after a large demonstration in Leamington the question was widely raised: 'Who shall arbitrate?'[48] But there was to be no arbitration. The farmers relying on the coming of harvest to abate agitation were mistaken. The traditional pattern of winter unrest was broken for the first time - the bitterness and extent of winter agitation had gone too deep and too far. The summer of 1872 was to mark the end of the agricultural workers' belief in the harvest as their main economic hope and their regard for the harvest as something other than mere work, shattered by the lengths that their masters were prepared to go to - even to threatening to let their crops perish rather than be dictated to, or allow the labourers the right to unite in their own interests. For the first time the labourers as a group had the courage and the conviction and some strength to challenge the farmers when they were most vulnerable. The Achilles heel of their masters, their dependence on a willing labour force at harvest time, had before been attacked only indirectly by the vulnerable individual labourer. This time it was different.

If all was quiet about Cubbington, Warwickshire, in early June[49] and a meeting at Eastleach described as 'a very tame affair',[50] labourers were on strike at Middleton Cheney, Buckinghamshire, in late June and a shepherd, milkman and other labourers were threatened for continuing work at Plummers Furze,[51] while at Brackely, Buckinghamshire, nearly 50 labourers had already formed a union branch on Monday June 10.[52] In the first week of July with men at Culworth 'out' three local farmers agreed to union terms while labourers on the Earl of Jersey's estate at Middleton Stoney, Oxfordshire, were receiving 18s a week in the hay harvest. At Banbury Cross the same week Arch was addressing an assembly of about 2,000, and at a meeting at Greatworth where most of the principal farmers were present and all the labourers from the surrounding district, 43 labourers joined the union. At Milton-under-Wychwood about the same number of people as at Banbury heard a long speech by Arch:

> Every working man ought to put his own value on his own labour, and the farmer had no business to say what we are worth, but he has paid us just what he liked. Who knows best the worth of labour, but the labouring man himself? Few farmers know what it is like to work from four to eight and feed flocks in snow, sleet and cold... think of the place of a working man and what can he do with 12s a week having a family of five or six to keep? Which works the hardest - the farmer or the working man? Rents are high because farmers compete against each other for the land (as many as 40 bidders)... they sell one another in the taking or buying of farms... and afterwards screw their profits out of the working man...[53]

Despite it being the middle of hay time, meetings continued to be held in different parts of the county, at Wolvercote, Eynsham, Beckley and Great Milton for example, and further membership was recruited.[54]

Such, however, was the labourers' mood of hostility and such the realisation that strike action alone was not going to alter the entrenched opposition of farmers that labourers were prepared to forego the prospect of harvest earnings and under union aegis migrate. At Wootton on Wednesday morning, 17 July about 40 strong able-bodied men mustered at six o'clock and proceeded with their wives and families to some iron works near Sheffield. The men, headed by a drum and fife Band, marched to Kirtlington Station.[55] On the same day at Gosford such was the proselytising zeal of a Kidlington man that he attempted to coerce 'using violence' a fellow villager to join the Warwickshire Labourers' Union and assaulted a farmer's son who objected to his interfering with other labourers.[56]

Migrations and strikes were to continue into the period of the corn harvest and events unprecedented at harvest time in the history of the countryside were to occur. On Wednesday 16 August

30 men left Middleton Cheney, and the rest of the labourers on strike went back to work at 24s and 25s a week, an advance of a shilling or two on the ordinary harvest wages.[57]

Farmers were able to read a general description of what was happening in the countryside and an assessment of the situation in the 'Agricultural Gazette' of 24 August 1872:

> the grain is laid in many places, and is beginning either to shed out or to sprout for want of reaping. Labourers in various parishes throughout the kingdom form themselves into processions, headed by bands of music, parade the public roads, march into the neighbouring towns and villages, and do no work. They demand high wages for the harvest, and in some instances stipulate for a fixed rate of pay during the winter months. Many farmers have acceded to their demands, so far as harvest operations are concerned, while others have threatened to let their crops perish on the ground rather than be dictated to by their men. In some places allotment grounds have been annexed to the neighbouring farms, and notice given to cottage owners to leave at the expiration of a certain time. Mowing and reaping machines have been useful during the present difficulty, but it happens unfortunately this season that grain crops in many places are flattened and tangled hence the machines are not so effective as they are in ordinary seasons, when grain is either erect or only leaning. It is painful to see 50 or 60 labourers parading the roads while the crops want harvesting. These men cannot exist long on the allowance of a trades' union. The winter is approaching, and then the pinch will come. This system of strikes has a tendency to harden the hearts of the well-to-do classes in matters of charity. Boards of Guardians will no doubt, during the next winter, look narrowly into the characters of applicants for relief. It is just possible that good cottage accommodation, and a little bit of land, added to a moderate increase of wages, would content the agricultural labourer.[58]

If the greatest single event in 1872 was the formation of Arch's Union, the two most significant occasions in the early stages of its campaign for better conditions were the calling in of the military to help in the harvest fields[59] and the imposition of prison sentences on the women of Wootton in Oxfordshire.[60] The solidarity, the strength, the resources and the determination of the farmers' opposition to change were clearly revealed. It was also clear that without a continuous organised challenge from the labourers a monopoly of rights would remain permanently with their masters and a maximum of duties imposed on the labourers. The phenomenal growth of unionism,[61] its attempts to raise wages even against ever-hardening opposition and its sponsoring of emigration did ensure a continuing struggle resulting in labourers' strikes and farmers' lockouts in the immediate years following 1872,[62] but the movement was to collapse through the circumstances of the depression. Ironically, a measure of independence which the farmers' opposition to unionism was implicitly out to prevent came to the labourers through the increasing opportunity to obtain allotments.

If the spearhead of unionism had come from the open villagers, this advantage came to them indirectly through its collapse in the depression years. Before 1872 the opposition to any united action had been immense; a significant delaying factor, however, had been the attitude of workers themselves. The hired men, the inhabitants of the closed villages, valued security of employment; the day labourers relied on their skills and initiative to compete for work against their fellows in winter and to bargain higher rates with the farmers in the peak periods of harvest employment; neither challenged circumstances except as individuals but both in their own way attempted to turn circumstances to their own advantage.[63] It is significant that though prosecution, under the Master and Servant Act, increased again during the three most active years of Arch's union, the number of offences committed were greater in the early 1860s when labourers were reacting to their own personal discontent in isolation.[64]

Concern for independence was in the main outside the experience of hired men:

> The first charge on the labourer's ten shillings was house rent. Most of the cottages belonged to small tradesmen in the market town, and the weekly rents ranged from one shilling to half-a-crown. Some labourers in other villages worked on farms or estates where they had their cottages rent free, but the hamlet people did not envy them, for 'Stands to reason' they said, 'they've allus got to do just what they be told, or out they goes, neck and crop, bag and baggage'. A shilling or even two shillings a week was not too much to pay for freedom to live and vote as they liked, and to go to church or chapel or neither as they preferred.[65]

The open villagers were committed to some independence; the increasing opportunity to obtain allotments made this independence more possible.[66] If unionism was dead by the late 1880s, the struggle for greater independence, the mood of resentment fostered by unionism and the rejection of subservience was carried over into their working lives in the knowledge that they were not totally committed to keep in the good graces of their employers. The lack of improvement in their condition in the 1860s demonstrated both the short-fall in their expectations and emphasised the limits that individual effort could achieve by accepting the circumscribed conditions of employment or the consequences of attempting to change them. Full employment did not often bring improved conditions to many closed villages. Charles Eyles, a farm labourer of Begbrooke, Oxfordshire, explained his circumstances in the late 1860s:

Table 8.4: Membership of the National Agricultural Workers' Union

Numbers [1]

	Berks	Bucks	Oxon
1874	2,000 (Reading)	2,050 (Wolverton)	3,000 (Oxford)
	2,750 (Wantage)	2,577 (Aylesbury)	2,599 (Banbury)
	4,750	4,627	5,599
1875	2,224 (Hungerford)	1,900 (Wolverton)	3,515 (Oxford)
	650 (W. Berks)	1,650 (Aylesbury)	2,300 (Banbury)
	2,874	3,550	5,815

Approximate Membership

1875	40,000	1880	20,000	1881-3	15,000	1884	18,000
1885	10,700	1886	10,366	1887	5,300	1888	4,660
1889	4,254	1890	8,500	1891	15,000	1892	15,000
1893	14,746	1894	1,100				

Funds (1889)

Income £3,018 4s 9d
Expenditure £5,173 12s 5d
Unemployment, travelling and emigration £132 12s 5d
Sickness, accident and funeral benefit £3,650 12s 0d
Funds at collapse £1,308 13s 1d

Note [1]: May 1873 - 71,835; May 1874 - 86,214; May 1875 - 58,652.
Sources: 'Labourers' Union Chronicle', 29 May 1875, p. 1, col. 3, p. 4, col. 1; J. P. D. Dunbabin, The Incidence and Organisation of the Agricultural Trade Unionism in the 1870s, 'Agric. Hist. Rev.', vol. 6 (1968), pp. 115-17, Table 1; A & P 1895 (c7900) XCII, Labour Department: Strikes and Lockouts, pp. 30-1; A & P 1890-1 XCII, Statistical Tables, pp. 96-9, 116-17.

> My cottage belongs to one of the landowners; I pay £3 a year for it and a small garden. The cottage is a very poor one with two bedrooms, one very small, both against the thatch. I get 11s a week all the year round; I'm a cattle man, and work on Sundays as well. All men here get 11s. I have no Michaelmas money. In harvest I have to go milking in the morning and at night, for which I get 6s a week, and I go reaping with my boy between times. Last harvest I earned 14s a week besides the 6s milking, but it was so excessive hot we could not work hard in the middle of the day; that extra harvest money is all the chance I have in a year of adding to the 11s a week. I have seven children. The eldest boy, 13, has 3s a week regular; he drives a team and does any odd job; the next boy, 11, is doing nothing. I can't get him anything to do at this time of year; the next two [eldest 9] are at school.

And for many 'open' villagers extra harvest earnings did not prove adequate to compensate for periods of unemployment which sometimes brought immediate hardship. Two labourers' wives living in Westcott Barton, an open village, complained: 'He has only done a day or two's work since harvest' (husband aged 21). 'My husband is out of work and has only done a month since harvest if it were all put together; it's almost starvation.'

Labourers in Steeple Barton were no better off: 'It ain't much living, the way we have to get on. Bread is about all we can get and about 2 lb of bacon for the family for a week. There is some people half starving in Barton. Some of them can't get work, and some won't work; but there's a good many as would work, if they could get it.'[67]

Despite the hired men having very real complaints, they made poor union material. Efforts were made to convert them to action but the very nature of their agreements, the close daily contact with their masters, their dependence on tied-cottage accommodation, made them wary of any direct confrontation. A notable effort to change this attitude of tacit acceptance was the publication, in 1884, of the 'Farm Labourers' Catechism', a biting parody with the opening lines:

Q. What is your name?
A. Clodhopper.
Q. Who gave you that name?
A. My masters, the landowners, and farmers when I was made a tiller of the soil, a scarer of birds, a keeper of cows and sheep, a producer of wealth, that my masters might live in idleness and luxuriousness all the days of their lives.[68]

However, in the last years of the century, and in a depressed agriculture with declining opportunities for even the regular and expected seasonal work,[69] they were prepared to challenge tacitly if not overtly the requirements and the assumptions of their masters. If the farmers had to cope with open revolt in the 1870s on the part of many labourers, a different problem faced them in the 1890s. With farmers' declining incomes they had to cope with their labourers' intransigence: 'Men want more looking after than they did', was the opinion of a Tetsworth farmer. 'They are more particular as to their time in threshing etc. Every year it gets worse. In fine weather they go on the allotments, and they want us to take them only when it is wet.' Similar opinions were expressed by a Kingston and a Thame farmer: 'They used to work ungrudgingly but now they will stop at the appointed hour, and put down a fork raised in the act of work. They will leave their horses with their harness on in the stable, and even refuse to give them water, if the clock has struck. They used not to get so drunk in their work as they do now. I have made up my mind to drop the beer...'; '[t]hey strive to do as little as possible and get off as soon as they can'.[70] Farmers resented the new independence of tone which had been increased by the possession of allotments at Shabbington and at other places, so that the men complained 'the farmers are all against us, they do not favour allotments, they will do all they can to prevent us getting allotments'. But farmers justified their resentment thus: 'there is a kind of independence about the men, you cannot find fault with them at all without their saying at once "well you can get somebody else for all I care": they have none of the old feeling for their employers' interest and they reserve all their best work for their allotments.'[71]

What was not realised was that without allotments, the pool of harvest labour despite the increasing use of machinery would have been further depleted; without the opposition to allotments there might have been more mutual appreciation. What also was not realised was that grudgingly given work at a grudging wage was grudgingly undertaken. It was the labourers' only weapon of attack and the only effective method of defence.

DISCORD

> It is often said that we paint arcadias that never did and never could exist on earth. To this I would answer that there are many such abodes in country places, if only our minds be such as to realise them. And above all let us be optimists in literature even though we may be pessimists in life. Let us have all that is joyous and bright in our books and leave the trials and failures for the realities of life... avoid as much as possible the painful side of human nature and the pains and penalties of human weakness.[72]

Writers have considered harvesting as a communal activity equating it with social harmony, inferring that if it existed at no other time, it was there in the harvest fields of the nineteenth century. Flora Thompson's villagers saw harvest time as 'a natural holiday':

> A hemmed hard worked 'un, the men would have said; but they all enjoyed the stir and excitement of getting in the crops and their own importance as skilled and trusted workers, with extra beer at the farmer's expense and extra harvest money to follow.[73]

But the harvest was also an occasion of drama, of distress, and even of tragedy, for it may be forgotten that the rural work situation could be as dangerous in its own way as work in any Victorian urban industry. If an awareness of this was not ever present, it was still sometimes expressed. The strain of those harvests of yesterday is brought home by the conversation of a Suffolk labourer's daughter.

> Father is very tired come bed-time [having worked] from 4 or 5 in the morning till 7 at night. If the weather is fine,

> uncle and them will finish this week, not father as he go by the week, and must do the stacks up and cover them. No accidents, thank God, but it have been hot for them all, and us as well...[74]

The arduous work in the summer heat indeed brought deaths from sun-stroke,[75] little children fell under and were killed by the wheels of the lumbering heavily laden carts and waggons, for not always could the young lads of slight years and sometimes little experience control the headstrong horses.[76] Older men like Henry Burton of Chalgrove who died on a Friday in September 1895 with 'Oh dear! Master Brown' on his lips as he placed the last sheaf on the last load of the corn harvest, and poor women like Sarah Lawrence aged 71 of Cholsey who returned home after the long toil of a harvest day in July 1869 to die in her bed,[77] were victims of exhaustion. There were acts of God when lightning struck[78] and sadly the acts of men when young girls left alone in the harvest field were raped.[79]

But the work situation itself was the main cause of accidents, often leading to serious injury and death. These accidents very largely happened to young boys and older and old men, and there seems to be a direct correlation between accidents and the kind of work performed. Extremely few happened to the women who were mostly engaged in tying sheaves and shocking the corn, and if cutting the corn, they would use the light and careful reaping hook and were therefore not engaged in accident-prone jobs. But the young boys, some barely 10 or 11 years old, were always put in charge of the horses, leading the waggons to and from the ricks and leading the horses when the mechanical reaper came into use. With comparatively little strength and lacking in experience, these lads could not always control heavy horses especially when suddenly frightened. It was often remarked at the time how well they coped: J. Arthur Gibbs in the harvest of 1896 watching two strong horses, tandem fashion, drawing a great yellow hay cart was anxious, for one small boy alone was leading the horses:

> Arriving at the ford, he jumps on the leader's back and rides him through. The horses strain and 'scaut' and the cart bumps over the deep ruts nearly upsetting. Luckily there is no accident. So much is entrusted to these little farm lads of scarce fifteen years of age it is a wonder they do the work so well.[80]

Some, too many, did not cope. George Henry Jones aged 10 of Abingdon, Charles Gardner aged 9 of Chimney, Joseph Smith aged 11 of Chipping Norton, Joseph Barrett aged 11 of Lathbury near Winslow, all leading waggons in the August of 1861, were all involved in accidents, and all of them died from their injuries.[81] Thomas Baylis aged 10 of Ilmington while driving some horses attached to a reaping machine was thrown among the corn by the plunging horses, startled by a gun shot, and his arm had to be amputated. This was in September 1873.[82] In 1875 a lad in Blenheim Park, a Poulton lad, and a 10 year-old Clanfield boy in 1890, all sustained serious injuries in the harvest field to limbs and thighs.[83] These again are but random instances; they were by no means exceptional.

The old and older men were equally vulnerable in their work which invariably consisted in unloading waggons and helping in the rick building. With deteriorating reflexes, often stiff and in pain with rheumatic joints, it was no wonder then that they fell climbing onto waggons, climbing ladders, and falling from these loaded waggons or from the ricks they were building.[84] At Great Tew near Deddington in July 1872 a 77-year-old agricultural labourer slipped off a waggon of hay, head downwards, and 'expired about 6 o'clock'.[85] At Deddington itself, four years later, an old man of 75, 'who being on a straw rick in a field called "Plank Piece" about 3 o'c... tumbled therefrom to the ground and thereby received such injuries that he died'.[86] On the 10 August the same year 1876 at Bampton, John Martin, a farm labourer aged 61, had just emptied a load at the rick and was standing on the raves of the cart when the horses suddenly moved: 'he fell to the ground striking his head against a ladder... died the same evening'.[87] At Murcot in July 1899 a 62-year-old labourer fell fatally, climbing down from a hay rick.[88] Falling from a straw rick, George Woodward, aged 56, a labourer of Appleton, died from haemorrhage of the brain at the end of harvest in 1898.[89] A shepherd, aged 50, employed by Mr Salmon of Luffield Abbey near Silverstone, Bucks, engaged with others building a rick in 1873 'overbalanced himself and fell from the top of the rick to the ground, a distance of 27 feet'. He died the next day from injuries to his spine.[90] An old man at Chalcombe slipped when climbing onto a waggon and died from his injuries; a 75-year-old Easington man collapsed and died climbing back to work on a rick; an aged man dislocated his neck from a fall off a wheat rick at Overthorpe; at Cornwell a labourer who fell backwards off a ladder while paring a rick, 'was a corpse in a few hours'. These fatalities all occurred in the last week in August 1847.[91]

Excessive drinking at harvest time was often the reason for much more serious consequences than a mere argument over who should pay for it though this was precisely the reason for a group of labourers appearing before the Aylesbury magistrates in July 1878.[92] Quarrels, fights, even serious injury - often arising out of trivial incidents - were part of the un-noted disharmony of harvest time.[93] A Fulbrook labourer found himself in prison for 14 days after an argument over their position on a rick with a youth aged 17 on 28 August 1872. He had struck the lad on the head with a prong for not getting 'out of the road' when told.[94] An assault on another labourer during a disturbance among labourers who quarrelled under the influence of the harvest beer cost a Hornton labourer 21s at Banbury Petty Sessions on Thursday, 8 September 1859.[95] After working at harvest cart until a late hour on the Maple Durham Estate in August 1856 a labourer was seriously cut by a fellow labourer on the knee and arm with a pea hook.[96] At Shipston-on-Stour Petty Sessions on 9 September 1865 John Simms of Brailes was charged with 'having on the 4th inst. assaulted John Henry Gregory of the same village'; both parties had been at harvest work for Mr Simpson of St Dennis Farm and had 'partaken too freely of cider'. Simms was convicted and fined 10s including costs.[97] At Deddington, in the harvest of 1873, a complaint of 'the beer being bad' caused a fight between farmer and labourer while they were 'pitching beans and vetches'. The labourer sued unsuccessfully for assault.[98] A dossier of such incidents could be collected from the local newspapers of the period; numberless unrecorded incidences must also have occurred, destroying some of the image of halcyon harvest days. If then drink was counter-productive, why was the alternative suggestion of the Temperance Movement virtually ignored? Very briefly the answer must be that beer and cider were part of the staple diet of rural workers; they provided not only sustenance but the necessary fluid that long days in the heat of summer demanded, and there was no cheap alternative freely available.[99] From the farmers' point of view it was an economical method of part-payment especially in cider districts. Without drink, work in the harvest fields would have been harder labour in nineteenth-century England. Without the consequences of excessive drinking the harsh realities of rural living would have been less harsh.

But whether because of drink or in the natural course of events injury and tragedy could result from social disorder. It may seem surprising that at harvest time anyone should contemplate suicide as did a young man aged 18 at Deddington in 1874;[100] but not so surprising that amity among the crowds of harvest workers was not consistently maintained. Harvesting was still essentially a communal activity and unlike working in the factories, where men faced machines, harvesters were reacting to changing conditions and each other from dawn to dusk. The use of sharp cutting tools, the lumbering horse-drawn heavily loaded carts and waggons, the tempo of the work, set by the incentive of increased earnings, for piece work or a harvest contract was most often the arrangement, all contributed to increase tension with the subsequent loss of critical care. Tempers could fly, the older men over-exert themselves in keeping up the pace, young children hinder as well as help, and the women grow tired and careless.[101] Harvest companies or individuals might be strangers to the neighbourhood and antagonism arise with the local workers. On 8 September 1853 Silas and Caroline Webb were fined 19s 6d for assaulting Joseph Grinstone in a Warborough field by throwing their hooks at him and injuring his head.[102] A dispute over choice of positions in binding the corn led to an assault on another labourer and cost a Southmoor man a fine of 1s and 11s 6d costs at Aylesbury County Petty Sessions in August 1899.[103] Two harvesters fought with reaping hook and reaping 'stick' over work in a field at Summertown in September 1874.[104] For 'unlawfully inflicting grievous bodily harm' to James Robinson with a reaping hook at Benson on 1 August the same year, a 30-year-old labourer was sent to prison for 'two calendar months with hard labour'.[105] At the end of harvest in 1848 on the Ditchley Estate near Enstone with 'harvest home' resounding through the woods, three lads engaged in an affray, 'influenced by beer and feelings of animosity', during which one was 'stabbed with a sheep knife and died'.[106] A mower, given unwelcome orders by a bailiff, 'threatened to stab him with [his] scythe... if he did not get out of the way'.[107] The Winslow magistrates in July 1856 fined him 10s and 9s costs for assault and being 'out of temper'.

It was not only cutting tools that became weapons in the harvest field. In the hay harvest of 1865 at Aston Magna near Campden during an altercation amongst the haymakers, a pitch fork was thrown which 'entered the eye and penetrated through the head' of one of the workers 'who [lingering] a short time only... expired'.[108] And work tools were not only weapons in the hands of men. A general quarrel took place among women working together at Letcombe Regis in August 1859; one of them was struck with a 'prong' (hay fork) receiving a wound in the arm. The culprit was fined 5s and 9s costs at Wantage Petty Sessions.[109]

If farmers viewed 'strangers' in the harvest field, whether 'companions of harvesters' or individuals or travelling Irishmen, as a very necessary pool of labour in the different conditions of commercial agriculture in the nineteenth century,[110] their presence was never very welcomed by local inhabitants. Hostility to any stranger was the perennial attitude and especially so at harvest time for higher harvest earnings were 'preserves' not to be lightly invaded and often defended. Apart, therefore, from disputes between farmers and their men, from labourers being sued for leaving work and labourers suing for non-payment of work performed, from accidents, from disputes leading to fights, injury, even death, from acts of God and the indecent acts of men, there was a structured hostility at harvest time which could be ever present when local inhabitants alone were rarely capable of supplying the total labour required for the safe gathering of the harvest. That it took positive shape and led to direct action was not surprising nor exceptional. Three Ipsden men brutally attacked a harvester hailing from Woodstock in August 1897.[111] In August 1876 John Clements of Saintsbury 'used violence' towards John Sanders ('a reaper and a stranger'), 'threatening and intimidating him with a view to coerce him and prevent him reaping [at] Weston Subedge... he was bound to give up his reaping and seek protection.'[112] A labourer dismissed from a Hardmead farm during the harvest of 1889 assaulted, 'thro' jealousy' and with the help of four others, his replacement ('being a Chicheley man').[113]

It often seems that work in the harvest fields meant for many writers

> Spreading the breathing harvest to the Sun
> That throws refreshful round a rural smell.[114]

As night approached they could think of

> [F]ew sights [as] more pleasing and exhilerating than the groups of reapers and mowers who are now to be met with in all the lanes and roads around a country village, just as the light fades into darkness, or gives way before the clear and mellowed lustre of the harvest-moon, returning merrily, if wearily home, after their long day's work. Their sunburnt faces still more highly coloured by heat, and it may be by beer likewise, wear a happy and good-humoured look at this season, which is not always to be found on them. They wish you good-night as they pass, in a franker and more friendly tone than usual. And these signs of human joy, combined with all the evidence of plenty lying round about one, enable a man, for the moment, to cheat himself into a real belief in the superiority of rural felicity.[115]

Even many labourers who had themselves worked in the harvest fields might in recollection have said 'T'were all mirth and jollity'.[116] But the evidence shows it was otherwise; not always, but sufficiently frequently not to be obliterated by the 'All things bright and beautiful' viewpoint of Victorian romantics and the rural writings of their successors reinforced by nostalgia.

NOTES

1 JOJ, 31 Aug. 1872. J. S. Wright, chairman of the Birmingham Liberal Association at Woodstock meeting Thursday, 29 Aug. 1872: 'They were the food producers... they ought to be better housed... fed... have more sufficient wages to educate their children.'
2 Mrs Parker, 'Supplement to Glossary of Words used in Oxfordshire' (1881), p. 90.
3 Reading Farm Records, BER 34.
4 Ibid., BER 21.
5 See, 'A Century of Agricultural Statistics' (HMSO, 1968), pp. 81-2, Table 36. Corn returns prices - compare top barley prices 11s 6d (1856); 11s 9d (1857); 12s (1868); 11s (1869). John Burnett, 'A History of the Cost of Living' (1969) p. 207. Average Gazette prices of British wheat - 14s 8d per quarter (1855), 64s 5d (1867), quoting W. T. Layton and G. Crowther, 'The Study of Prices', 3rd edn (1938), p. 234.
6 Burnett, 'A History of the Cost of Living', p. 199; 'Cost of Living Index', 1286 (1840), 969 (1850), 1314 (1860), 1241 (1870), 1174 (1880), based on Seven Centuries of the Price of Consumables Compared with Builders' Wage Rates, 'Economica', vol. 23, no. 92 (1956); E. Victor Morgan, 'The Study of Prices and the Value of Money' (Historical Association, 1950), p. 24; 'From 1850 to 1873 there was a rise... of about 20 per cent.'
7 'Folk-lore', vol. 24 (1913), p. 76.
8 'Ag. Gaz.', 30 Mar. 1867, p. 327.

9 A. H. Ruegg KC, 'The Laws Regulating the Relation of Employer and Workman in England' (1905), pp. 17-18.
10 PP 1875 XXX, RC on Labour Laws 2nd and Final Report, pp. 8-9; for chronology and details of acts relating to hiring of labour as further outlined in text, see pp. 9-11.
11 See Table 8.1 for convictions, p. 125.
12 Frederic Harrison, Imprisonment for Breach of Contract of the Master and Servant Act, 'Tracts for Trade Unionists', no. 1 (1868), reprinted as 'Year of the Union', Edmund Frow and Michael Katanka (eds.) (1968); for Convictions under Sec. 14, see Table 8.2, p. 127.
13 'Windsor and Eton Express', 24 July 1875. For subsequent convictions, see Table 8.3, p. 128.
14 See 'Aylesbury News', 22 August 1846. Ardley man charged at Bicester Petty Sessions on 14 August with having misconducted himself by getting drunk on Sundays - engaged for 30 weeks to Michaelmas at 6s per week and 30s Michaelmas money.
15 PP 1893-4 c. (6895-II) XXX, RC on Labour: The Agricultural Labourer, B - Thame, p. 210 and p. 206.
16 See Table 8.3, p. 128.
17 JOJ, 11 Sept. 1875.
18 Ibid., 15 Aug. 1875.
19 Ibid., 21 July 1860.
20 Cf. 'RLSC', 20 Jan. 1872. The Staffordshire Agricultural Labourers' Protection Society demanded: 3s 6d per 12 hours work and 3d per hour overtime for labourers, 3d for 9 hours work and 4d per hour afterwards.
21 JOJ, 11 Sept. 1875.
22 'Reading Mercury', 21 Aug. 1875, p. 8.
23 OT, 1 Oct. 1859.
24 Ibid., 4 Sept. 1847.
25 'Bicester Herald', 3 Sept. 1875; see also ibid., 3 Oct. 1873. Two labourers break contract to cut wheat at 14s per acre.
26 JOJ, 11 Sept. 1875.
27 'Bucks Herald', 9 Aug. 1856: note current bricklayer's wage 24s per week, ibid., 2 Aug. 1856.
28 'Aylesbury News', 4 Jan. 1840.
29 JOJ, 13 Aug. 1859.
30 'Reading Mercury', 28 Aug. 1875.
31 OT, 30 Aug. 1890.
32 The Employers and Workers Act, 1875 (38 & 39 Vic. c.90), pt. III, 10.
33 See PP 1893-4 c. (6894-II) XXXV, p. 211. But cf. 'Reading Mercury', 12 Aug. 1882: at Reading County Court, 10 August, farm foreman dismissed in July (18s per week and £5 Michaelmas bonus) 'claimed for part thereof' - 'Not entitled to claim'; also PP 1893-4 XXXV B-II Wantage, p. 219, 'Michaelmas Money... cottage rent free both... forfeited'.
34 PP 1893-4 (C6894) XXXV, 'the men appear to be afraid to do it and dread the expense'.

35 JOJ, 17 Sept. 1859.
36 Ibid., 1 Oct. 1859.
37 Ibid., 5 Aug. 1865.
38 OT, 11 Oct. 1890.
39 'Bucks Herald', 2 Nov. 1889.
40 JOJ, 19 Oct. 1861.
41 Ibid., 9 Aug. 1862.
42 OT, 17 Aug. 1889.
43 RLSC, 17 Feb. 1872.
44 See ibid., 24 Feb. 1872 and ibid., 9 Mar. 1872.
45 Ibid., 23 Mar. 1872.
46 See ibid., 30 Mar. 1872.
47 Ibid., p. 9, special edition, The Agitation Among the Farm Labourers; ibid., 16 Mar. 1872, leading article, The Land and the Labourers; ibid., letters, A Labourer on the Strike, from an agricultural labourer of 35 years experience, The Labourer on Strike, from a yeoman of the old school.
48 Ibid., 6 Apr. 1872, leader, Who Shall Arbitrate?
49 Ibid., 8 June 1872. An enterprising journalist sought a labourers' meeting in vain.
50 JOJ, 22 June 1872.
51 Ibid., 29 June 1872.
52 Ibid., 22 June 1872.
53 Ibid., 6 July 1872.
54 'Oxford Chronicle', 9 Sept. 1882. Compare Headington Church of England Temperance Society, monthly meeting, Tuesday 1 September, 'The attendance was not large, owing to work in the harvest fields and St. Giles' Fair.'
55 JOJ, 20 July 1872.
56 Ibid., 27 July 1872.
57 Ibid., 25 Aug. 1872.
58 'Ag. Gaz.', 24 Aug. 1872, pp. 1144-5; 'Labourers' Union Chronicle', 7 June 1873: charity withdrawn at Hardwick, Bucks - '12s enough to live on'.
59 JOJ, 24 Aug. 1872. Labourers' meeting held at Wootton to discuss soldiers sent from Aldershot to get in the harvest; ibid., 17 Aug. 1872. Farmers about Farrington, Berks, had also recruited soldiers from Aldershot 'to take the place of the regular labour'.
60 E.g. Reading Farm Records, OXF 14/1/1, f.3, rise of 3s per week, 9½ hours day, 8½ hours Saturday demanded of Richard Holton Lamb by Sibford Branch of the National Agricultural Labourers' Union, March 1873 (f.4); notice to leave employment 14 Apr. 1873 by eight employees, 'George Lines his mark X' etc., 'Bucks Advertiser' 13 Sept. 1873; 202 Buckinghamshire men and women arrived in Brisbane on 30 June - 'a very healthy and superior class of labourers' (Brisbane Courier', 8 July 1873); see also Frederick Clifford, 'The Agricultural Lock-out of 1874' (Edinburgh, 1875); A. Clayden, 'The Revolt of the Field' (1874); F. G. Heath, 'The English Peasantry' (1874), pp. 187-202;

S. E. Fussell, 'From Tolpuddle to T.U.C.' (Slough, 1948), pp. 59 -85 - Arch still claimed 60,000 members in April 1875 (p. 85); R Groves, 'Sharpen the Sickle' (1949), Chs. 3 and 4.

63 'Folk-lore', vol. XXIV (1913), p. 76. 'Self first, then your turn next best friend' (Oxfordshire saying).

64 See Table 8.2, p. 127.

65 Flora Thompson, 'Larkrise to Candleford' (1965 edn), p. 6.

66 Note: The depression made illogical the farmers' objection to the provision of allotments, but they still attempted to prevent it, e.g. 'Bucks Herald', 10 Aug. 1889, p. 3, col. 2, R. J. Treadwell, farmer, led objection to the provision of allotments at Rural Sanitary Authority Meeting at Twyford, Bucks.

67 PP 1868-9 (4202-1) XIII, App. B, p. 570 (5b) (4b) (4c) and (3).

68 Bodleian MS 23214e 11/6, 'The Farm Labourers' Catechism' (Harvey Reform Printing Works, Andover, 1884).

69 PP 1893-4 XXXC, BII Wantage, p. 218 (11), 'the land is terribly starved here, full of couch grass quitch etc., and the result is it grown four quarters of wheat instead of six or seven' (at East Hendred). 'Farmers scarcely employ one man to 100 acres on an average, and six men are employed where 10 ought to be' (at Hampstead Norris).

70 Ibid., Bl Thame, p. 207, III (18).

71 Ibid., p. 210, X (68).

72 J. Arthur Gibbs, 'A Cotswold Village' (1898), Preface, p. vi.

73 Thompson, 'Larkrise to Candleford', p. 255.

74 A. Jobson, 'Suffolk Yesterdays' (1944), p. 60.

75 'Windsor and Eton Express', 18 July 1868.

76 'Oxford Chronicle', 10 Sept. 1869. 'Girl killed by wheel of barley cart... playing with other children... horse led by boy of 11. Coroner noted this very improper practice.'

77 Ibid., 21 Sept. 1895; ibid., 3 July 1869.

78 'Warwick Advertizer', 4 Aug. 1860. Man loading waggon killed by lightning, also three horses at Slough, Bucks, 29 July.

79 OT, 7 Sept. 1889. Girl, aged 8, drawing bonds for harvester alone in barley field - assaulted at Wallingford.

80 Gibbs, 'A Cotswold Village', p. 369.

81 JOJ, 24 Aug. 1861; ibid., 14 Sept. 1861; ibid., 21 Sept. 1861.

82 Ibid., 27 Sept. 1873.

83 Ibid., 11 Sept. 1875; ibid., 14 Aug. 1875; OT, 20 Sept. 1890.

84 See Heath, 'The English Peasantry', p. 224. 'Oxfordshire Labourers about Woodstock... poor rheumatic creatures dragging their unwilling limbs over the stones'. And, quoting Dr Watt of Witney, 'Children are employed too young in heavy ploughed land; it tells on them later in life; when they get about fifty they go at the knees.'

85 JOJ, 3 Aug. 1872.
86 Supplement to JOJ, 30 Sept. 1876.
87 JOJ, 19 Aug. 1876.
88 Ibid., 22 July 1899.
89 Ibid., 1 Oct. 1898.
90 Ibid., 18 Oct. 1873.
91 Ibid., 28 Aug. 1847.
92 'Bucks Herald', 13 July 1878. The labourers declared they were drunk.
93 'Reading Mercury', 14 Oct. 1882. A labourer harvesting with three others would not 'stand' a quart of beer, assaulted the ring-leader and was sent to prison for one month.
94 JOJ, 21 Sept. 1872.
95 Ibid., 17 Sept. 1859.
96 Ibid., 21 Aug. 1895.
97 'Bucks Herald', 9 Aug. 1856.
98 Ibid., 4 Oct. 1873.
99 Cf. 'Folk-lore', vol. XXIV (1913), p. 76. 'Bread is the stuff of life, but beer's life itself' (Oxfordshire saying).
100 JOJ, 29 Aug. 1874. Thomas West having taken more beer than was 'good for him... threw himself into a mill stream with the intention of drowning himself'.
101 Ibid., 19 Aug. 1865, reaper aged 48 complained of 'a pain at his heart... obliged to leave work... died in a few hours at Buckingham'; 'Bucks Herald', 23 Aug. 1890, 'lad of twelve cut off his little finger... with a sickle at Oakley'; ibid., 'shepherd threatened his wife with a hook... quarrel over boy making "bonds" inefficiently'; OT, 30 Aug. 1890, 'pregnant wife helping her husband in the harvest at Stanton St. John poisoned by fungi... died'.
102 'Oxford Chronicle', 3 Sept. 1853.
103 JOJ, 19 Aug. 1899.
104 OT, 14 Sept. 1874.
105 'Bicester Herald', 23 Oct. 1874.
106 'Banbury Guardian', Sept. 1848.
107 'Bucks Herald', 26 July 1856.
108 JOJ, 22 July 1865.
109 Ibid., 3 Sept. 1859.
110 Cf. 'Victoria County History of Berkshire' (1906), vol. 1, p. 183, 'there were generally bands of men wandering about to seek extra work'. But note the different attitude of farmers in past times, e.g. 'Attempts to check the practice of having men from outside to the disadvantage of the residents is shown by a by-law passed by all tenants of Bright Walton in 1330 to the effect that no strangers should be received in Autumn corn harvest and that no-one belong to the liberty should work outside without a licence.' (Ct. R. ptfo 150 no. 69 Bright Walton). Note also the need for specific laws to obtain harvest labour. But once hired, either annually or specifically for the harvest, labourers were

under the compulsion of the 'labour laws'; farmers could and did invoke the Master and Servant Act et al. to retain their labour.

111 OT, 13 Aug. 1897.

112 JOJ, 19 Aug. 1876.

113 'Bucks Herald', 6 Aug. 1889.

114 Edward Jesse, 'Favorite Haunts and Rural Studies ' (1847), p. 16, quoting James Thompson in a description of eighteenth-century hay making at Ritchings Park, Bucks - 'a host of mowers and haymakers'.

115 Anon, Harvest, 'Cornhill Magazine', vol. 12 (Dec. 1865), p. 359.

116 Gibbs, 'A Cotswold Village', pp. 387-8. 'Fifty years ago... there was four feasts in the year for us folk' (John Brown, aged 78).

9

HARVEST CUSTOMS

> When the harvest is over, you should sell the gleaning, or pull the stalks yourself; or else, if the ears be few and labour dear, they should be eaten down. For you must look to the main chance lest in this matter the cost exceed the return.
>
> M. Terenti Varronis, 'Rerum Rusticarium Libri Tres', trans. Lloyd Storr-Best (1912), Bk. 1, Ch. LIII, The Gleaning, p. 105.

If, at the end of harvest, gleaners took over the fields gathering every grain left lying in the stubble, it was not always without opposition. It is perhaps difficult to realise that this custom was a matter of serious contention in the last century when only the purely operational consequences of the use of the combine-harvester, involving field threshing and the burning of straw on the stubble, so recently brought it to an end. It is also difficult to appreciate that so many days of backaching work was thought worth the effort, but to the gleaners the gathered grain represented one of the mainstays of the home - a safeguard against winter's privations.[1] This practice, indeed, was seen as a common right of the countryside embodying the Mosaic Law, that the gleaning should be left 'unto the poor and the strangers'. Despite the decision in the Court of Common Pleas against gleaners,[2] so rooted was this custom that the gleaners continued to collect the fallen grain despite farmers' objections and the suggestion of some writers that the practice ended in 1788. In a period when ownership of property was sacrosanct and the game laws so rigidly enforced that a tenant farmer could be prosecuted for shooting rooks in his own field,[3] this annual invasion by the rural community of private property must be seen as a manifestation of the belief that right of access to the soil was a fundamental right which should not and could not be revoked. Magistrate courts throughout the nineteenth century confirmed the ruling that 'No person has at common law the right to glean in the harvest field. Neither have the poor of the parish legally settled [as such] any such rights,'[4] but, despite this, the practice continued into the twentieth century. In the 1880s in North-East Oxfordshire:

> [A]fter the harvest had been carried from the fields, the women and children swarmed over the stubble picking up the ears of wheat the horse rake had missed... It was hard work from as soon as possible after daybreak until night-fall... At the end of the fortnight or three weeks that the leasing lasted the corn would be thrashed out at home and sent to the miller who paid himself for grinding by taking toll of the flour...[5]

A decade before, due west a few miles into Warwickshire at Tysoe, in the 1870s, as soon as a field had been cleared,

> [E]ach woman with children she had been able to bring took a 'land' or ploughing ridge, laid out a sheet with a stone at each corner, and then the whole company began to work slowly up the ridges, all the figures bending hands deep in the stubble, 'leasing' fallen ears. Each gleaner had a linsy-woolsey bag hanging from her waist. Tiny boys and girls had tiny bags. Long straggling straws were gathered, while broken off ears were dropped into the bags.[6]

J. C. Cornish relates that at Debenham, Suffolk, in the 1860s:

> Farmers were seldom so niggardly as to send a drag rake over the stubbles after the sheaves had been carried; and the gleaners might begin to gather stray ears as soon as the waggons were loaded. But there was one rule for them to keep - they must not go before the waggons for a man with a wide wooden rake went along where the shocks stood and swept up the stalks remaining there.[7]

Walter Rose describes how at Haddenham, Bucks, in the 1870s gleaning was the means whereby almost everyone could obtain a small stock of flour:

> the women and children would sally forth at morn to the fields to return at even with neat bundles of wheat gleaned from the fileds. This, in the evening or on wet days, was rubbed or beaten with sticks. Corn would be taken to the windmill to be ground into flour.

The record of a widow's gleaning is worthy of mention. 'She with the help of her three sons, gleaned one harvest time the total of twelve bushels of wheat, five of barley and five of beans. Rising at 4 a.m. she returned at 7 a.m. with her first large bundle of gleaned corn, then after breakfast, her children would accompany her and help her throughout the day until late at eve.'[8]

Against a background of continuing gleaning and a continuing series of prosecutions it was natural that some attempt should be made to have the practice redefined as a 'legal' practice. It is understandable that these pleas failed when the summing up in

1788 against Houghton and his wife, who claimed the right as legally settled poor, is examined. Already in 1787 the claim of a right to glean 'indefinitely by the poor, necessitous, and indigent persons' had been denied on the grounds that it was: 'inconsistent with the nature of property which imports exclusive enjoyment'; 'Destructive of the peace and good order of society, and amounting to a general vagrancy'; '[I]ncapable of enjoyment, since nothing which is not inexhaustible, like a perennial stream can be capable of universal promiscuous enjoyment'. But the rebuttal of Houghton's claim was put in much more comprehensive terms and with much more weight by Lord Loughborough, the Lord Chief Justice, and J. Heath and J. Wilson, the two other sitting judges:

> Although it is insisted on, that this custom... is coeval with the constitution... yet the first mention of it is in the Trials per Pais, a more extrajudicial opinion of Lord Chief Justice Hale, 'That by the custom of England the poor have a right to glean'. The next author, Lord Chief Baron Gilbert, ... says that the poor are 'allowed to glean', which implies... permission, rather than a right... It has been argued... that no corn is claimed but what is abandoned by the owner; as if the owner had cast it from him, and it became the property of the poor by... a sort of occupancy. [But] no property can be lost by abandonment, for the owner may at any time resume the possession... Such a custom as will support the plea, must be universal and everywhere the same, otherwise it is void for its uncertainty... it is partial and no part of the general customs of the realm. From the best enquiries, I find that this custom is not universal. In some counties, it is exercised as a general right, in others, it prevails only in common fields, and not in enclosures, in others it is precarious, and at the will of the occupier. In the county [Timworth, Suffolk] where this action was brought, it never in practice extended to barley; nor is the time ascertained. In some counties the poor glean whilst the corn is on the ground; here the usage is laid to be after the crop is harvested.
>
> The practice of gleaning was originally eleemosynary. But it is the wise policy of the law, not to construe acts of charity, though continued and repeated for never so many years, in such a manner as to make them the foundations of legal obligation.
>
> Wherever there is a right, the law provides a remedy, if that right be obstructed. But suppose the owner of a field were to set fire to the stubble, or to flood it, and prevent the poor from gleaning, what remedy could they have?... Every institution which is to be found in the law of Moses was not enforced by the judge, many of them being left to the consciences of men with temporal blessings on those who observed them. The right of gleaning is given by the same law as well to the 'stranger' as the 'fatherless and poor'. We

have already infringed it, as we have decided that the stranger has no right to glean in the case of Worlledge v Manning.

The law of Moses is not obligatory on us. It is indeed agreeable to Christian charity and common humanity, that the rich should provide for the impotent poor; but the mode of such provision must be of positive institution. We have established a nobler fund. We have pledged all the landed property of the kingdom for the maintenance of the poor, who have in some instances exhausted the source.

It would open a door to fraud, because the labourers would be tempted to scatter the corn in order to make a better gleaning for their wives, children and neighbours. It would encourage endless disputes between the occupiers of land and the gleaner. It would raise the insolence of the poor, and leave the farmer without redress... It has been alleged... that the poor ought to have a share of benefit... it may be answered that they receive from the advanced price of labour, a recompense in proportion to their industry... [Lord Loughborough].[9]

...[a]s there is no evidence of this custom of gleaning prevailing uniformly throughout the kingdom, as the practice of it is uncertain and precarious, and as it would be attended with great public inconvenience, if it were enforced as a right... the plea is therefore bad... [J. Heath].[10]

...No right can exist at common law, unless both the subject of it, and they who claim it, are certain. In this case both are uncertain. The subject is the scattered corn which the farmer chooses to leave on the ground, the quantity depends entirely on his pleasure. The soil is his, the culture is his, the seed his, and in natural justice his also are the profits. Though his conscience may direct him to leave something for the poor, the law does not oblige him to leave anything. The subject then is uncertain and precarious.

...[i]f there be a right, there must also be a remedy if that right be infringed... if a farmer were to give permission to his brother, or friend of another parish to glean his fields, the poor of his own parish could have no remedy in law, for what they might think a prior right [J. Wilson].[11]

Why then was gleaning allowed to continue? There were many reasons, not least that the ruling of 1788 did not specifically prevent gleaning. If farmers acquiesced in the practice there were no problems; if they sought to control gleaning for social or economic reasons they now had a legal backing, but only if they invoked it. In a period of distress in 1802 when 46 persons at Kidlington were in receipt of parish relief, the only action to prevent gleaning because the practice was being abused was to publish printed handbills against it.[12] With a custom hitherto so general and accepted many must have been unaware that the 'right' had been extinguished, that a privilege could not be taken for granted, that farmers might legally object. In October 1889 four Kidlington villagers were indicted and tried at Oxford County Court for gleaning corn against the will of the occupier of the land:

> As it appeared to the Court that the parties were not aware they were committing a felonious act, they strongly recommended an acquittal: but they were clearly and decidedly of the opinion that no person has a legal right to lease without the consent of the occupier of the land; that the practice of leasing is an indulgence founded on benevolence, and ought when permitted, to be exercised with gratitude and respect. That leasing should never be attempted or permitted before the bulk of the corn is carried from the ground: and that then it would be highly discreditable to any person to endeavour to prevent or interrupt it.[13]

It is clear from the magistrate's final comments that there was no general desire to obliterate the ancient custom but only to ensure that the initiative of allowing the practice to continue was known to rest solely with landlord and tenant farmer. That the issue of gleaning was still a very live one in 1887 is made clear by the editor of the 'Bicester Herald' deeming it a matter of moment to reproduce an extract from an article on gleaning previously published in 'Farm and Home':

> There is a belief amongst many of the poor that gleaning is a right to which they are legally entitled; but this is an error. The custom is an ancient and very common one, so that the privilege is scarcely ever withheld. Poor people look upon gleaning as a means of providing a stock of flour for winter use, and preventing them from obtaining this help for a family means the infliction of a hardship upon the class. Abuses of the permission are, however, met with, and this tends to unpleasant wranglings. One way to meet the difficulty is to insist firmly that all corn shall be out of the field before gleaners enter, and then allow them full liberty before animals are turned in. A rule of this kind on a farm is understood in the locality, and if the gleaners find they have a fair chance, they will not overstep the privilege.[14]

The numerous court cases throughout the century demonstrate how widespread gleaning remained. They also illustrate the considerations that led to farmers objecting to the practice. It seems clear from the almost complete absence of cases involving wheat that there was rarely any objection to gleaning this crop. With barley and beans it was quite otherwise for the main body of cases almost always seem to relate to them. There can be various explanations for this situation. Against the background of a formal illegality and with the increasing emphasis on commercial profitability in farming practice there was still no economic reason to forbid the practice of gleaning wheat - in fact a sound reason for allowing it since in a four- or five-course crop rotation gleaning could be considered a 'cleaning' operation in preparation for winter ploughing. A further consideration was that the gleaned corn went to make bread for the villagers' own direct consumption and not to maintain small livestock where feeding after the harvest supply from gleaning was exhausted might lead to subsequent pilfering from the farm granaries. The reverse situation operated in respect of barley and beans for both were mainly livestock food, especially for pigs. If gleaning was allowed, it might encourage labourers to keep stock; if fed in situ, turning pigs into the bean and barley stubble, the economic advantage remained with the farmers. Hired labourers living in farm cottages were, in fact, often prohibited from keeping either pigs or poultry. Thus if gleaning other than the 'bread' grain had been permitted, further trouble might have arisen, for the unrestricted open villagers would have obtained a positive advantage denied to the hired men. The risk of inviting future pilfering and the economic sacrifice of their own interest was reason enough for farmers refusing the privilege of gleaning these crops. In addition, if the barley crop had been undersown, turning cattle or sheep in to graze immediately after carting provided very useful grass keep which imminent autumn frosts might quickly spoil.

If gleaning wheat was generally allowed and gleaning barley and beans generally disallowed there was still a further restriction which amounted not to prohibition but to selectivity on the farmers' part, for quite often the privilege of gleaning was reserved for those engaged in the actual harvest work itself on individual farms. Thus villagers could glean only in the fields where they had actually harvested. All others were excluded or only allowed to glean when the farmers' labourers were free to glean themselves.[15] 'Why will you not allow the poor to lease your fields?' a Dinton farmer was asked when prosecuting a labourer attempting to glean in a field already carted and carried in the harvest of 1840. 'I thought my own labourers should have the first chance of picking up what corn might be left on the ground, and they were carrying barley at the time. If they could have gone, I should not have objected...' The labourer received a firm reprimand: 'You know that leasing is robbery [without consent]', but no other penalty.[16] When actual prohibition was not the case, the main restriction however was the insistence that no gleaning should start until a field was completely cleared, for both the temptation to rob the sheaves and the risk of loss was otherwise too great. In 1853 a Dorchester farmer was induced to print a notice warning gleaners not to enter his fields until they were cleared of their crops, because of the 'loss and annoyance to the farmers of the neighbourhood' which their methods of gleaning had caused. The notice had little effect and he was charged with assaulting a woman who had refused to give up the barley she had gleaned. He was acquitted as he 'had acted in defence of his right' but could obtain no redress for his 'stolen' barley unless he was prepared to bring an action of trespass.[17] In 1859 a Brill farmer prosecuted a labourer for continuing to glean in his barley stubble after having been warned. He had brought extra pigs to pick up barley and beans. The case was dismissed with a caution.[18] In 1862 five Blackwell women were charged with stealing one gallon of barley, value 6d, at Shipston-on-Stour. They claimed they had permission to glean, but they had to pay a fine of 2s each. The fine was unusual but the gleaning took place in an open field where some barley was still uncarted and 'as the law in open field farming is very stringent' the magistrates felt they could not be lenient.[19] In 1890 at Sutton Courtney two boys had to pay the harsh fine of 5s each for gleaning beans after they had been previously cautioned. The farmer explained, 'We allow them to glean Wheat, but then they are on sufferance...' but 'not beans or barley'.[20]

It might be said that the custom of gleaning survived because of the ambivalence on the part of all the rural protagonists. With no sufficient economic reason to object, the farmers were prepared at the worst to ignore the annual invasion of their fields. The magistrates, while insisting on the illegality of gleaning without the tacit or overt consent of landowners, were not prepared to see the offence as much more than a very minor misdemeanor. The labourers themselves, while continuing to insist on a blanket right to glean, were prepared to accept the implicit denial of this right by accepting restricted gleaning when it was to their advantage, e.g. when gleaning was restricted to the actual harvesters on a particular farm.[21]

If farmers imposed restrictions on who should be allowed to glean and what fields should be gleaned, the villagers themselves voluntarily agreed to the times when they could glean. This was sometimes imposed by the ringing of the gleaners' bell which was rung in the parishes to denote the time to start and to finish gleaning for the day so that all could share fairly. In 1860 at Aldeby, a village in Norfolk, but within three miles of Beccles in Suffolk, and at Tibenham no one was allowed to commence gleaning before the morning ringing or to remain in the fields after that in the evening, though it was not considered a usual Norfolk custom. At the same period it was reported as being observed in the parish of Churchdown and in the neighbouring parish of Sandhurst in Gloucesterhsire until 'within the last few years'.

> In Churchdown, which is a large corn-growing parish, the gleaners' bell became mute when the parish was enclosed under the Enclosure Act, about a dozen years ago... In the case of the gleaners' bell, the 'enclosing' was not always its quietus; for in Sandhurst, which is still unenclosed, the farmers stopped it a few years ago, perhaps from hearing that their neighbours had done so. In Sandhurst the bell was rung at six a.m. and eight p.m., the gleaners themselves paying the clerk for his trouble. In Churchdown the same, except that the clerk was paid ten shillings by the parishioners in vestry.[22]

But at Tadmarton and Swalcliffe near Banbury and in some Oxfordshire parishes the custom still prevailed and the gleaners were 'very particular in attending to its warnings'.[23] At Driffield in Yorkshire, although the 'harvest bell' had been rung from time immemorial at five o'clock each morning and at seven each evening to warn the labourers in the harvest fields when to begin and cease their labour, the gleaners' bell was specially introduced in some corn villages in the 1840s. The Driffield harvest bell or 'barley bell' was still ringing at the same times in 1901 but there is no longer any specific mention of the gleaners' bell.[24] A. H. Cocks in fact stated in 1897 that 'so far as my information goes this [the gleaners' bell] is no longer rung anywhere in the country',[25] though William Andrew, writing in the same year, was not so emphatic. 'In some of the more remote villages of the country, the gleaners' bell is [still] rung as a signal to commence gleaning.'[26] As late, indeed, as 1911 at Upton St Lawrence in Gloucestershire a bell was rung after harvest as early as 5 a.m. to let people know they might go to glean or lease; 'many would be ready to start'.[27] Indications therefore suggest that the custom survived fairly late in the nineteenth century. At Aston Abbots in Buckinghamshire, where it was also called the leasing bell, it was rung at 8 a.m. and 7 p.m. and was only discontinued about 1883. At Olney it appears only to have rung for 30 or 40 years and was discontinued in 1885 or 1886; here it rang at 7 a.m. and 7 p.m. At Ravenstone it was discontinued about 1854 while at Sherington it was rung 'formerley'; at Grandborough it was 'said' to have been rung formerly.[28]

A recognition of the acceptance of gleaning, but emphasising its restrictive right, was the farmers' participation in the practice by leaving a sheaf in the field. Until this 'guard' sheaf, sometimes called a 'policeman', was removed no one was allowed to glean.[29] At the turn of the century everyone at Garsington gleaned, but only wheat. 'If someone went to glean the barley stubble - you still wouldn't stop them?' 'Oh no', said a local farmer, 'the country people in those days, they knew when to go, when not to go.'[30] A colourful illustration of how seriously country people took the practice of gleaning was the revival at Rempstone in Nottinghamshire in 1860 of an old custom of proclaiming a 'Queen of the Gleaners':

> The village crier, having 'proclaimed the Queen' nearly 100 gleaners assembled at the end of the village. Women with their infant charges, boys with green boughs, and girls with flowers, the whole wearing gleaning-pockets; children's carriages and wheelbarrows, dressed in green and laden with babies etc. were in requisition... [A] royal salute was shouted by the boys, and the crown brought out of its temporary depository. This part of the regalia was of simple make; its basis consisting of straw-coloured cloth, surrounded with wheat, barley, and oats of the present year. A streamer of straw-coloured ribbon, dependant on a bow at

> the crown, hung loosely down; a leaf of laurel was placed in front, while arching over the whole was a branch of jessamine... The ceremony of crowning was now performed; after which the Queen, enthroned in an arm-chair decorated with flowers and branches, moved... [to] the 'first field to be gleaned'.

At the scene of labour her proclamation speech was read:

> You have made me Queen of the Gleaners till the harvest is finished. I will try to rule by right and in kindness, and I trust to your obedience that I may not have to exercise my power. I will now tell you my laws, which shall farther be made known by the crier of the village.
> 1st. My attendant shall ring a bell each morning, when there are fields to be gleaned.
> 2nd. Half-past 8 o'clock shall be the hour of gleaning, at the end of the village, and I will then accompany you to the field.
> 3rdly. Should any of my subjects enter an ungleaned field without being led by me, their corn will be forfeited and it will be bestowed.

Then following a brief address from the Queen of the Gleaners and a suitable song, the 'whole commenced their labours in the barley field of Mr. James Moor, the first field to be gleaned'. In the corn-growing counties further south it might have been exceptional if only because gleaning barley was being allowed.[31]

The practice of ringing bells to denote the time of day was an essential need in the countryside until the introduction of cheap pocket watches late in the nineteenth century. A 'dinner bell' was still being rung in 1897 every week day at Elmberton, Sherington and Winslow, and at Olney in Buckinghamshire.[32] As late as 1900 the rent of a piece of land in Hanney Parish near Wantage called the 'Bellman's Swathe' was paid for a bell to be rung at 5 a.m. from 19 September till 2 February to call agricultural lads up to attend to their horses.[33] Although it seems that the 'gleaners' bell' was a specific custom there might otherwise have been some confusion about which bells served what precise function. Alfred Gatty stated in 1848 that the bells that were rung at 'six in the morning and at noon' calling the 'labourer to his daily toil and to his dinner' were thought to be continuations of the mediaeval Ave Maria bell which tolled at these hours for prayer.[34] But in the summer months, even outside the period of harvest, labourers worked longer hours, for 'dawn to dusk' had a nice flexibility and was inevitably a longer period than in the winter months. Hired men as distinct from the day labourers, whose summer working days were extended at the end of the day, commenced work in winter usually at 6 a.m., in summer at 5 a.m. For example at Newport Pagnell as late as 1887, when it was temporarily discontinued, the fifth bell was rung every week day at 5 a.m. from

1 March to 1 November.[35] Thus it is probable that the 'harvest bell' was in fact the regular bell rung during the rest of the year to tell workers the time, and only at harvest time was it specifically termed the 'harvest bell', for then it was of greater moment for many more workers to heed its ringing. Rung at seven in the evening it would indicate the end of the harvest day for day workers, or later in the century the start of overtime payment. But if it was the custom to ring a bell at 8 p.m. it was specifically the curfew bell, a relic of the custom said to be instituted by William the Conqueror to advise everyone to extinguish light and fires to prevent the conflagration of houses.[36]

Whether there was an absolute right to glean or not, belief in this right persisted with perhaps an exaggerated emphasis on the claim because of the serious economic consequences to labouring families if the custom had ceased. In the past it would seem not to have been an unrestricted right and was indeed the subject of innumberable village by-laws. At Islip on 18 June 1462, such a by-law was formed which stipulated that 'no tenant or anyone else [should] glean corn on the arable after it [had] been cut, or pea pods, with permission; [on] pain of forfeiting the corn so gleaned and of paying 12s to the lord for each offence'.[37] This did not imply that gleaning was in jeopardy, however, merely that it came under the accepted control which pertained to all agricultural activites in open-field cultivation; the time to plough, to sow, to gather the harvest were all regulated - so also the gleaning. Theoretically if not in practice, all gained from or all suffered under these rules of husbandry. In the individualistic conditions of the nineteenth century, the only way to preserve the practice of gleaning, previously the concern of an integrated community, was to insist that it persisted as an individual right. Of the implications of constituting it a legal right, the Court of Common Pleas was fully aware when making its decision. Permitting it, however, as a privilege in a spirit of paternalism avoided the distress which prohibition would bring upon the labourers, yet maintained intact both property rights and the existing social structure.

A letter in 'The Times' of Saturday, 29 August 1848 headed 'Let the Poor Glean' made it a live issue in these terms:

> In many counties, on one excuse or another, this so called privilege has been either restricted... or only permitted at all when the grasping owner or occupier has with his horse rake picked up almost every ear of corn... or [turned] his pigs in to feed on the droppings of the harvest field. Were I asked to name one single boon which more than another at once tend to win back their love for those above them, I would say restore to them at this very time in all its fullness, the privileges to glean with their families as they used in former years.[38]

Though it remained a live issue, for once social and economic attitudes did not coincide and gleaning continued. It can be seen as the last formalised survival of peasant rights in the nineteenth century: formalised because there were customs associated with gleaning which rural society as a whole, clergy, laity and labourers, gave conscious or unconscious recognition to and in which they were active participants. The Court of Common Pleas did not silence the gleaners' bell; farmers did not cease to leave a last shock of corn in the field to be removed as a signal for gleaning to commence. Even when these outward witnesses of an ancient customary 'right' were gradually abandoned, no countryman thought that gleaning would cease. First, because it provided a valuable extra to supplement the labourers' larder; second, because it was good farming practice. No amount of raking could clear the broken ears of corn; no farmer wanted sprouting wheat in his subsequent crop, be it barley or oats or beans or peas, or more probably in his root crop.[39]

HARVEST CELEBRATIONS

> Harvest-home was the most important and the best-loved festival. The whole of the toilers - men, women and children - attended this; the brewhouse and kitchen, too, were needed to hold the company. Besides the farm hands, the blacksmiths and wheelwrights came, invited by the farmer. The parson frequently attended, and the village constable managed to creep in and mingle with the rest.[40]

The 'harvest home' was part of the fabric of eighteenth-century rural existence. Alfred Williams, however, was right to emphasise its continuing importance in the nineteenth century, for in a changing order it still, if only momentarily, restored the dying or already moribund attitudes of a peasant past. Farmers and labourers were united not only in celebrating the successful conclusion of a task of unprecedented magnitude but, as agriculturists, under the same roof again; eating and drinking at the same table. The growing division between 'master and servant' was 'held'; indeed for one day at least the roles were reversed. Consuming the farmer's food and drink, in kind, quality and quantity not perhaps to be tasted again until the end of another harvest, denoted equality; served by the master, prepared and served by his wife, it signified something more, a recognition of their importance.[41]

At Long Handborough ('Amborough) in the 1870s, the supper consisted of large joints of meat, plum pudding and 'tins of potatoes, all baked in the brick-oven, with an unlimited supply of beer'.[42] An 85-year-old Henley carter had fond recollections (in 1914) of the old harvest feast:

> Thor' was master an' missis a-hippin' an't up in front on us, an' owl' Moll Fry a-skippatin' about wi' the cups. Presently

all an us fell to. 'Lar father, byent 'o gwain to saay graace', missis ses. Then maaster stood up, shet 'is eyes, an' put 'is 'ands together an' said: 'O Lord make us able
To aat all on the table'.
'Oh, dad, you wicked fella' cried missis, an' all the young ums baaled out 'Amen'. The paarson purty nigh chokked 'isself wi' a lump of bif an' 'ed to wesh it down wi' a cup o' ale, an' when tha sung 'Drenk, bwoys, drenk', a sed 'twas no need to tell 'em that for tha could do that very well wi'out being telled.[43]

Among the songs sung in Oxfordshire was a refrain beginning:

I love a shilling, a jolly, jolly shilling,
I love a shilling, as I love my life;
A penny I will spend, and a penny I will lend
And ten pence carry home to my wife.

It continued 'I love a ten pence' and ended 'And nothing to carry home to my wife.' The shilling may well have been the harvest 'earnest' shilling. Another song, 'Oliver Cromwell lies dead in his grave' may well have implied that no puritanical restrictions were going to interfere with everyone's enjoyment. A toasting song, 'Here's a health unto our master', with minor variations according to county and dialect, was sung all over the country including lines to 'our mistoris' (Oxon).[44] Another drinking song reported as peculiar to Dorset was:

As I was a-riding over a mountain so high,
I saw a pretty girl that plea-sed my eye;
She plea-sed me eye, but pla-gued my heart.
From this cup of liquor we never will part.
'Twill do us no good, 'twill do us no harm.
Here's a health to my love, over left arm, over left arm
Here's a health, etc.

This accompanied the custom of 'drinking to your love over the left arm'. Each man was obliged to drain his mug or horn-cup of ale, by passing it outside of and over his left arm, which would be thrown against his chest. 'Great merriment was afforded when some of the older hands, through age or other infirmities, failed to accomplish this in a satisfactory manner.'[45] A very popular festive song about Bampton was the 'Cuckoo's Nest':

It is known oh my darling it is no such thing
Pay an commar sense may tall yar it is a great sin
Yar maidenhead to lose and then to be abused
And have no men a daisy with your cuckoo's nest.
Some like a girl that is pretty in the face
Some like a girl that is slender in the waist
But give to me the girl that can wriggle and will twist
And at the bottom of her belly is the cuckoo's nest.[46]

In Oxfordshire if a female was asked to sing but was not anxious to do so, she would decline by singing a formal refusal beginning, 'You have asked me to sing' out of tune,[47] but in Sussex when a labourer was unable to respond to a call for a song he usually rendered the following recitation:

Bell rings, Up goes I
'Betty' says he 'Sir' says I
'Now Betty you may breakfast along with me'
'La, Sir, I couldn't think of such a thing'
'But Betty' says he 'you must'
So I breakfasted with master all the time mussus was at Bath.

Other incidents of everyday life were then gone through, the last line being, 'And in the middle of the night I dreamed my soul was carried up to heaven in a hand-basket'.[48]

It must be assumed that many of the songs were bawdy and unprintable for harvest homes were not held in disrepute for drunkenness alone. 'The Harvest Songster' though published by a street ballad printer in the early years of the century was highly respectable, containing verse such as:

As they followed the last team of Harvest along
And end all their toils with a dance and a song

from 'The Last Team of Harvest' and

See the loaded harvest wain
Bringing home the ripen'd grain

from 'The Harvest Wain'.[49] These songs can be seen as the secular precursor of those hymns composed later in the century specifically for the Church Harvest Festival. But although all so well known now, harvest hymns such as 'Come ye thankful people come', 'Lord of the harvest once again', 'Summer scenes are glowing', 'Fair waved the golden corn', 'We plough the fields and scattered' and 'Once more the joy of Harvest' were thought worth collecting together with some others and publishing in a booklet 'Special Harvest Hymns' in 1901, so recently had they been composed.[50]

Preceding the supper there were customary practices that took place in the harvest fields. Two at least were widespread and have been widely described - one associated with the cutting of the 'last sheaf', the other with the safe carrying of the 'last load'. At Ducklington in Oxfordshire it was still the custom during the 1830s when the last load was to be carted

to send down to the field a number of band-boxes containing women's dresses and a good deal of finery for the men and horses. Four young men then dressed themselves up, two to represent women, and they sat in couples on the four horses that drew the load... and on reaching the house were treated to cakes, etc.[51]

At Long Handborough, labourers riding home on the last load used to shout

Hip, hip, hip harvest home
A good plum pudding and a bacon bone
And that's a very good harvest home.[52]

In Berkshire, with as many labourers as possible riding on top, the last load was drawn to the rickyard accompanied by this song repeated at short intervals:

Well ploughed well zawed
Well ripped well mawed
Narra load owverdrawed
Whoop whoop whoop whoop harvest whoam.[53]

At Wendlesbury, Oxfordshire, in the earlier years of the century the labourers crowding the top of the waggon knew what to expect when they passed through the village singing

Harvest home! Harvest home!
We wants water and cant get none.

From every house buckets of water were thrown on them.[54] Laying an ambuscade for the 'hock' cart was certainly an eighteenth-century custom in Buckinghamshire, but as late as 1872 it was witnessed in the county of Rutland. On this occasion, the load was decorated with green boughs and in addition to pitchers full of water being dashed over it, the driver and occupants were pelted with a shower of apples.[55] The custom of decorating (exclusively with ash boughs) the harvest cart was also still extant in various parts of the midlands in the 1860s.[56]

Depending on the locality, the last sheaf was sometimes made into a 'Kirn Dolly' or a 'Corn Baby' or a 'Corn Maiden' or simply termed the 'Harvest Baby'. If a girl harvester was delegated to cut the last sheaf she was called the 'Har'st' or Harvest Queen, and, decked with summer flowers, rode on the last load. Otherwise the last sheaf, made into an effigy, occupied this place of honour. Both were seen as representing the Corn Goddess, Ceres.[57] The customs carried out at the actual cutting of the last sheaf were variously called 'To Cry the Mare' in Hertfordshire, 'Calling the Neck' or 'Crying the Knack' in Devonshire and Cornwall, 'Cutting the Frog' in Worcestershire, 'Shouting a Kirn' in Northumberland, 'Cutting the Kirn or the Queen' in Berwickshire, 'Winning the Churn' in Ulster and 'Cutting the Calacht' in Eire.[58] Just as various were individual local differences.

As practised in Hertfordshire this custom 'To Cry the Mare' was essentially pre-nineteenth century. The reapers tied together the tops of the last blades, stood at some distance and, throwing their sickles, attempted to cut the knot. The winner highreaped the sheaf and retained 'the prize with acclamation and good cheer', i.e. the grain.[59] A variation in some midland counties which survived into the early decades of the nineteenth century was for the successful reaper to cry

> 'I have her!' 'What have you' the others cried out. 'A mare!' he replied. 'What will you do with her', was then asked. 'Send her to ———' naming some neighbouring farmer whose harvest work was not completed...[60]

In 'Calling the Neck' a bundle of the best wheat was selected. The reapers crowded round, took off their hats, bowed to the 'neck' and began a long harmonious shout. 'The Neck' repeated three times was followed by 'Wei you', 'Way you' thrice repeated whereupon the first reaper to seize the 'neck' ran to the farmhouse which was guarded by a dairy maid armed with a bucket of water. If he gained entry unobserved he demanded tribute from the dairymaid, otherwise he received a drenching.[61] 'Cutting the Frog' as the custom was called in Hamilton Kingsford, Worcestershire, and still followed as late as the 1850s, appears to have been used in two senses: for cutting up to the last stalks or for cutting the last stalks when they had been plaited together. Both were considered an honour but it was equally uncertain which reaper would gain either or both honours, for as the cutting progressed the reapers changed places after each 'drift' or 'bout' and it could not be predicted to whose lot it would fall to cut up to the last corn.[62] In Berwickshire in 'Cutting the Kirn' the outcome was made still more uncertain for the reapers were blindfolded before throwing their reaping hooks. As a reward for this extra difficulty the successful competitor was tossed up in the air 'three times thrice' on the arms of his brother harvesters.[63]

Frazer in the 'Golden Bough' stressed the universality of all harvest customs, their link with ancient fertility rites and how dominant was the direct connection between the animal kingdom and vegetation in peasant societies.[64] Mannhardt specifically centred his view on fertility demons linked to the last sheaf. When the last sheaf was treated in various ways, these were different methods of making use of the vegetation spirit. In antiquity the vegetation spirit inherent in the last sheaf grew into the goddess Dometer. Von Sydow,[65] however, suggested that these spirits were pedagogical fictions; that if dangerous beings were said to be in the corn to prevent children trampling it down, and when it was cut they did not appear, elders might say as in Russia that the beings were previously 'as high as the corn' but, after harvest, 'no higher than the stubble'. Or the customs were causal fictions; the 'Crying of the Neck' had no mysterious roots in past or present belief, but served to let neighbours know that the harvest cutting was completed.[66] But whether the customs are to be seen as survivals or only as formalised 'chance pranks' the advent of the mechanical reaper meant that those associated with the 'last sheaf' had to disappear.

The most significant influence in putting an end to harvest customs in general must be taken to be a decline of a communal spirit in the gathering of the harvest. Where these customs survived longest - in the northern counties, in the west of England[67] and in eastern counties such as Suffolk - either the harvest was 'taken' by specific groups of immigrant harvesters, or labourers agreed to receive each an equal sum for harvest work, or payment was not mainly in cash but included the provision by farmers of both food and drink.[68] In these districts, therefore, there was either a special group identity or some definite attachment to the harvest on a particular farm. By singling out the last sheaf for special attention, the importance of this task was emphasised; any rivalry between harvesters which could be disruptive while the cutting was unfinished could be switched to the time when cutting was completed, any inter-group rivalry was fixed on striving to finish first; on whose farm this might be was left in no doubt for the customary loud proclamation made the whole neighbourhood aware.

An explicit example of this communal participation was the fact that in Yorkshire two harvest suppers were provided. Atkinson was puzzled that there should be two terms for the harvest supper and sought an etymological explanation.[69] But the 'kirn' supper occurred when corn cutting was completed and was given to those harvesters or bands of migrant harvesters who then departed to cut corn on another farm, while the 'mell' supper was the traditional 'end of harvest' celebration for all the workers and helpers who saw the harvest completed. This end of harvest celebration was called the 'horkey' in the Fens and East Anglia but here again must not be confused with the 'frolic', a celebration held by the labourers themselves and organised at the beginning of harvest. The money came either out of an advance made by the farmer, to be deducted out of the agreed price for the harvest contract before work actually began, or from the 'largesse' collected on their first day of harvest work when, leaving work early, they resorted to the nearest ale-house to fortify themselves with a bout of hard drinking. This 'frolic' was often considered an essential preliminary enabling them to face the long arduous days of harvest work ahead.

The harvesters' custom of demanding 'largesse', that is gifts of money from passers-by or visitors to the harvest fields, could however be a daily occurrence throughout harvest. It was peculiarly an eastern counties custom, still extant in the 1880s, although the ritual of 'hallering largesse' was then considered abandoned.[70] While this sometimes occurred at the end of the day in the corn fields, the definitive occasion was at the end of harvest when the labourers paraded the village. After collecting donations they grouped themselves in a circle with the 'lord' in the centre; when he gave the signal, 'largesse' was shouted as loudly as possible dwelling on the first syllable lar-r-r-r. This was repeated three times, concluding with three yells and 'Thank you, Sir!' if any donor was present. They then retired to the village. This 'frolic', (largesse spending) was a communal merry-making - especially when a 'Horkey' was no longer provided by the farmers, even if the ritual thanking was abandoned.[71]

It was not therefore the mechanical reaper, in counties such as Berkshire, Buckinghamshire and Oxfordshire which was mainly responsible for curtailing old customs, indeed its adoption was late in comparison with the northern counties. Rather it was that the practice of individual piece-work payment made harvest work essentially competitive in the early stages of harvest, while the old peasant community identity with the harvest on a particular farm or estate was undermined by the casual link that 'open' villagers had with farmers in their locality. In a competitive wage situation the labourer had to be more concerned with his own or his family's earning power than with preserving communal customs; farmers, too, were less inclined to tolerate peasant practices if their main concern was profit. The chronology of persistence or decline, however, is difficult to establish since the majority of writers have concentrated more on the form and content of individual regional customs than on recording the period or even (except very broadly) the places in which they occurred. But it is necessary to establish some time scale to appreciate the interaction between harvest customs, economic forces and other social factors. Such an attempt will therefore be made in regard to one of the more important and least regional of these customs - the harvest home supper.

'HARVEST HOME' INTO HARVEST FESTIVAL

> The harvest-men sing summer out
> With thankful song and joyous shout
> And, when September comes, they hail
> The Autumn with flapping flail.[72]

Although August was seen traditionally as the harvest month and Hone as late as 1827 could say 'Harvest-home is the great August festival of the country',[73] it could not remain so in the nineteenth century with the greatly increased corn production. The harvest, in fact, was to extend backward to late July and forward to late September, or even exceptionally into October depending on local and seasonal variations. The fact that Lommas-Day,[74] a calendar feast of thanksgiving, fell on 1 August (Old Lommas, 12 August) suggests that with pre-nineteenth-century yields and the predominant use of the sickle, so that grain was safely gathered leaving a high stubble to be mown later, the harvest was customarily completed by the end of August. It also suggests that the 'good old English custom' of celebrating the ingathering of the crops with a harvest home, still heavily documented for the earlier years of the nineteenth century, was not a religious festival but the end of harvest that gave rise to convivial rejoicing. Leigh Hunt speaking of the eighteenth century describes 'Harvest-home' as still the greatest holiday in England: 'and it concludes at once the most laborious and most lucrative of the farmer's employment and unites repose and profit'.[75]

Indeed, E. Chambers links the harvest home not to Christian traditions but to the pagan rites of the Romans:

> In Romulus's year, December was the tenth month whence the name viz from decem ten; for the Romans began their year in March... At the latter end of this month they had the Juveniles ludi and the country people... the feast of the goddess Vacuna in the fields, having then gathered in their fruits and sown their corn; whence seems to be derived our popular festival called harvest-home.[76]

The interesting fact, however, is that not only did the time of celebration change - from the end of August to even as late as late October - but the form of celebration changed in the nineteenth century. From the mediaeval Lammas celebration of the first communion in the bread of the new corn,[77] there is a transition to a purely secular harvest home celebration, compared by Jefferies perhaps over-luridly to the Greek Saturnalia,[78] a custom which continued well into the nineteenth century. By the 1860s the secular custom was dying and a religious celebration, 'The Harvest Festival', had once more become firmly established. While the eighteenth-century Church cared little or nothing for the harvest,[79] by the end of the nineteenth century the Harvest Festival was firmly established in the Church calendar. '[There] is scarcely a church in England in which a Harvest Thanksgiving Service is not held', wrote S. Baring Gould in 1876.[80] Thirty years before there was scarcely a clergyman in the land who had thought of such a possibility.

How did this transformation come about involving as it did a change both of attitude in the clergy and of custom amongst agriculturists? It was a conjunction of two separate streams of thinking, the one economic on the part of the farmers, the other social on the part of the clergy, that effected this change in a period when not only the sheer physical scale of the harvest was having an impact but the endless debates on the Corn Laws, the general unrest due to Chartism and the serious economic effects of a poor harvest on the lives of so many no longer cushioned by the old Poor Law, could not be ignored. Indeed, when in 1838 the Bishop of Hereford asked Melbourne for a harvest festival he refused on the grounds that 'one would be obliged to have a thanksgiving for everything'. 'In the starving year of 1842 the abundant harvest saved lives and public authority [did] issue a form of thanksgiving.'[81]

The exact date of the first Harvest Festival is in doubt. The generally held view is that 'we owe our Thanksgiving Service to Hawker of Morwenstow' in 1846.[82] According to the 'Peterborough

Advertiser' of 3 October 1886, however, '[T]hese seasonal festivals... are said to owe their origin about forty years ago to a former Rector of Elton, the late Bishop P. Claughton', but Cuthbert Bede a prolific contributor to 'Notes and Queries' refutes this:

> From 1850 I... frequently visited Mr. Piers Claughton, but I do not remember anything about his harvest festivals... I have now before me the manuscript of a paper on Harvest Festivals... I read... on Sept 6 1861. It is evident from details in that paper that such festivals were by no means general... were only then being introduced... I have notes of that date of many such festivals - at Patshulls and Belbroughton... and in numerous other villages and towns... but not... prior to 1860.[83]

It is clear from a tract 'The Harvest Festival at Lilbrook', written in 1861 by an anonymous vicar 'believing that it contains some useful hints for his brother clergy who may contemplate substituting a Church Festival for the wretched harvest-suppers still continued in many parishes',[84] that such festivals did occur before 1860, though the first authorised one in Ely Cathedral was held in October 1861. It was not until 1863 that any official directive came from the Province of Canterbury. When Convocation assembled in July a letter from Sir George Grey was read stating that the preparation of a thanksgiving service for harvest by permission of the Queen in Council as proposed was of 'so doubtful a nature that the Lord Chancellor thought Her Majesty ought not to comply with the application'. Therefore a decision was taken that 'each clergyman was now to arrange the details of his own harvest festival guided by... the diocesan Church Calendar'.[85]

It must be assumed, therefore, that a gathering groundswell of opinion prior to 1860 with a number of individual isolated parishes already conducting Harvest Festivals, had by the early 1860s gained sufficient support to obtain general approval from the Church hierarchy. In Oxfordshire the first Harvest Thanksgiving Service at Kencott was held in 1858.[86] In Buckinghamshire it was in 1860 that a newspaper report informed its readers that '[T]he Waddesdon Harvest Festival this year produced £31.8s to the funds of the Bucks Infirmary', suggesting that it had not been the first,[87] though at Wing only the third Church Festival had occurred by 1873. This had been held in the National School where 'the whole of the labourers were given a holiday for the greater part of the day, without deduction for loss of time'; about 200 attended the dinner in the afternoon with a service in the evening.[88] But by 1863 in Berkshire, the gentry, clergy and local farmers had combined so successfully that separate celebrations on the neighbouring farms about Swallowfield Park were abandoned in favour of not merely a religious thanksgiving, but an expression of national loyalty. That year the lord of the manor and the local gentry led a procession which included the local band to church for a morning service at which Rev. Charles Kingsley preached. Three hundred and forty labourers and their wives were given a dinner at midday in place of their usual harvest supper; the afternoon was spent in games and unspecified rural pursuits; and to the satisfaction of the organisers, no cases of drunkenness were reported.[89]

Doubtless there could have been few of the clergy who would have disagreed with one suggested reason for holding such festivals:

> Were these parochial festivals more general throughout our country parishes, I do believe that they would in no small measure tend to elevate the country labourer by causing him to feel that he had those over him who cared both for his pleasure and his good.[90]

The general concern of the Church at the supposed and indeed actual drunkenness and lasciviousness that harvest home occasioned, made it inevitable that there would be an intervention, and a sustained interest in replacing secular occasions of festivity at harvest time by a more seemly and controllable form of celebration.[91]

For economic reasons the seeds of the disappearance of the harvest supper were there in the early years of the century. Payment in kind, an essential feature of pre-industrial agriculture, persisted throughout the nineteenth century since its method presented advantages when hard cash was the least readily available commodity and the Truck Acts did not apply to agriculture.[92] One of these customary payments was the harvest supper, and the provision of meat and drink in lavish quantities to ill-fed labourers was manifestly not economically advantageous. Dunkin saw the 'harvest home' as a relic of servile anomaly in a wage economy with the reservation that 'it remains for the farmers to consider how the prospects of merry making stimulate the exertion of the workmen'.[93] Arthur Young in 1805, concerned only with the economics of profitable farming, was already substituting custom for convenience and economic gain by paying 5s or 7s 6d in lieu of 'earnest', dinner, gloves and 'hawky' or harvest supper when making his own harvest agreements, and advising others to do likewise.[94] In the 1850s a Tangly labourer remembered his employer saying to the men 'What shall I give you instead of a harvest home?' They got 3s 6d apiece. On the Ditchley Estate 2s 6d was the amount paid in 1878, the first year that the harvest was not celebrated with a 'harvest home'.[95] This sum was the generally accepted commutation irrespective of a rise or fall in wages, and pro rata payments of 1s and 6d were paid to women and boys.[96] On a Cambridgeshire farm in 1893 written into the 'harvest contract' (the practice of taking the whole of the harvest by 'companies of harvesters' most common in East Anglia) was the sum of 2s 6d or 'horkey', the East Anglian term for the harvest supper unknown elsewhere.[97] In the event the men each received their money payment for no supper was provided. But it was the willingness of the farmers to co-operate with the clergy that decided the gradual dominance of the Harvest Festival, though this was not always forthcoming. A parochial harvest home festival was held at Kings Sutton in 1862:

> In the morning the leading farmers of the parish with their labourers marched in procession to the church headed by the Kings Sutton band... then paraded the village to the Bell Inn where all... partook together of a substantial dinner of English fare. [Responding to the toast of the 'Bishop and Clergy'] the vicar remarked [on] the conduct of a few persons in the parish, who gave their men half-a-crown each, but did not allow them to attend that festival.

It was early days yet. Around the same time, for example, at Henley on Thames, 'one of those good old-fashioned meetings, so universal in the time of the past generation, was held at Stonor Farm where Mr. Jas. Wells assembled his labourers, together with those who had assisted in gathering in the "produce of the season" numbering upwards of 170 persons'.[98] If there were objections on the part of farmers to co-operating with the clergy, though prepared to abandon the harvest supper, the labourers in their turn objected to losing their individual celebration, though if compensated in cash they would, as one contemporary commentator in 1881 put it, 'have a merry-making and go to the ale-house'.[99] Some therefore were sorry to see the end of the harvest home for the good that stemmed from the farmer 'condescending' to preside at the 'supper' table was lost while the drunkenness and horse play at the end of the harvest was not prevented. Indeed there were others who objected when harvest homes were beginning to be discontinued. One farmer's wife was very angry: 'They had always had one in her father's time and she didn't see why they shouldn't have one always, so she and her son provided a feast for the men down in the village.'[100]

Besides the clergy's approval, however, there must have been general social support for the new forms of harvest celebration, especially among hospital administrators, for it became customary to donate the Harvest Church Service collection to the local hospital. An interesting and revealing exception was the donation of upwards of £33,000 over a period of six years during the agricultural distress from 1887 to the Royal Agricultural Benevolent Institution which gave help to distressed farmers.[101]

Customs die and new ones take their place. It is ironical that at the same time as industrial dependence on the harvest and urban concern for the countryside declined, and as agricultural depression fell on English arable farming, it had become the established custom to conduct a Harvest Festival Service in not only rural but urban England. And those rural workers who finally turned their backs on harvest fields in ever-growing numbers found the corn they had abandoned, because no living could be obtained from harvesting it, cut and bound in sheaves decorating the churches in the urban parishes where they settled.[102] But customs though replaced have a habit of lingering on. The exceptional Harvest Festival of the late 1850s became customary over the next 40 years. The harvest home became exceptional but did not disappear until the twentieth century. In the southern parts of England it perhaps disappeared faster but a Shippon farmer was 'keeping up the good customs' in his home barn in 1872.[103] On a farm at Long Crendon about 70 labourers ate their harvest home supper in the same year.[104] A substantial supper was enjoyed by 150 harvesters at Fawley Court in 1882.[105] In the same year a magic lantern display terminated a pleasant evening at the Benham Estate in Letcombe Regis.[106] In 1889 the harvest home was spent in a 'convivial manner' at Hyrons Farm, Amersham.[107] In 1893 there was a harvest home at Barton Stacey, and about 140 people sat down to a harvest supper at Sherfield Court, Sherfield-on-Loddon where 'songs were sung maintaining the good old custom'.[108] It needed maintaining to compete with the ever-spreading harvest festivals. In Berkshire in this same year, in this same week, harvest festivals were being held at Crowthorne, Woodely, Easthampstead, East Hendred; the following week at Goring, Mortimer, Benrow and Silchester to mention only a few.[109]

In 1898 those who had assisted at the 'ingathering of the harvest', about 40 in all, were given a harvest supper in the good old English style at Park Farm, Bersleigh, and at Noke the end of harvest was celebrated in a similar fashion, although at Chasewoods Farm a harvest home attended by 30 guests was considered a unique affair for at Hailey there had been 'none in the last twenty years'. In the early 1890s the harvest home at Deanery Farm, Clanfield, was reported as 'one of those old fashioned customs which is now seldom heard of' with 64 persons attending the supper; that at Signett near Bampton, where about 50 were present in a large granary, as an 'old English custom so fast dying out'.[110] In 1882 at Swallowfield Park the form of the new 'custom' had infiltrated the old; 'dinner' was at 2 o'clock in the afternoon and a cricket match was played.[111] At New Lodge, Windsor, in 1893 a cricket match between estate workers and the farm labourers was also played but this preceded the supper after which 'the rest of the evening was given up to mirth and song' ending however at the early hour of 10.30 p.m. with the 'happy gathering' singing 'God save the Queen'.[112] This occasion could hardly be compared with the harvest home at Tangley barely four decades before which did not finish till five o'clock the next morning. 'The master couldn't get up' and the carter having to plough next morning went to fetch his horses out, but fell on his face in the litter.[113] But compared with the harvest home at Purley, Surrey, a week later which commenced with a service and then a cricket match, supper at five with tea but no beer, followed by Evensong at 7 p.m., the celebration at Windsor must have been quite boisterous.[114]

Whether in a 'bastard' form or as a revival the harvest home was to survive the century. At Garford near Kingston Bagpuize it was reported as 'still kept up' in 1889 though in many parishes 'dying out';[115] in 1892 80 of its villagers were still celebrating 'in the olden style'.[116] At Great Rollright in 1880 it was 'revived' with much spirit having 'somewhat fallen into abeyance'.[117] In 1898 the number of people in the neighbourhood of Burford who had a harvest home could be 'counted on the fingers of one hand',[118] but at Middleton Stoney, a year later, it had not fallen into disuse and was still regarded as a general holiday among the villagers.[119] 'Jackson's Oxford Journal' in this last year of the old century still considered the old custom important enough to devote to it a leading article commenting on the increase in harvest festivals and the decline of the harvest homes but not their demise.[120]

J. Arthur Gibbs at the close of the century noted how the custom of the harvest supper was dying:

> Altho' many of the farmers here have given up treating men to a spread after the harvest is gathered in, there is still a certain amount of rejoicing. The villagers have a little money over from the extra pay during the harvest, so that the gipsies do not do badly by going round the village at this time. The village churches are decorated in a very delightful manner for these feasts, such huge apples, carrots and turnips in the windows and strewn about in odd places; lots of golden barley all around the pulpit and font; and perhaps there will be bunches of grapes, such as grow wild on the cottage walls, hung round the pulpit.[121]

But even as the harvest supper was being superseded, and on the farms where it was still customary, due rather to revivalism than to unbroken tradition, the harvest thanksgiving service itself was considered by some to be changing. William Andrews in 1892, while suggesting it was useless to regret past practices, did not share Gibbs's delight or decorative taste and was regretting the increasing lavishness and lack of taste in the outward forms associated with Church celebrations:

> Even the harvest-thanksgiving service, with its accompanying cereal and horticultural decorations of church and chapel seems destined to change. The decorations are too often overdone. We have seen in some churches piles of fruit and vegetables that would furnish a shop, in addition to sheaves of corn and stacks of quartern loaves. In some instances, a more deplorable display has been made in the shape of a model of a farmyard.

To what degree unionism accelerated the decline could be a subject of further research, but certainly hostility between labourers and farmers over wages and conditions was not conducive to the continuance of a custom based on a harmony of interest. 'It [the harvest supper] promoted a good feeling... Now alas! it is far otherwise... thanks to the modern agitator' was one contemporary observation made in 1881.[122] Yet a certain ambivalence is suggested by the situation at Weston Colville, Cambridgeshire, at the height of union activity: 'one farmer sacked all his men but would take them back' if they ceased to be union men, another provided the customary harvest supper.[123] Already by the early 1870s, however, with farmers influenced by economic and clerical persuasion, the harvest supper was half-way to extinction. The harvest celebrations that occurred in a South Dorset village in 1873 might be taken as providing an almost exact reflection of the current attitude, not only to the customary provision of a harvest supper but to all traditional customs associated with the end of harvest. While the church was decorated and a service held in the afternoon attended by all the labourers in the village, the entrance gates of the principle farms were likewise decorated with an arch of evergreen, flowers,corn, etc., crowned with a sickle and a scythe, and each individual farmer provided his own labourers with a harvest supper. While customs in the fields were no longer engaged in, the ritual of 'Crying the Knack' was performed first by men and then by women, but only as part of the 'after supper' entertainment.

The abandonment of the harvest home meant the final obliteration of the last links with a peasant society. The division between farmer and labourer was already manifest in the widening economic gap between labourers and even the smallest tenant farmer; the social gap became fixed when they lost this last opportunity to eat under the same roof celebrating, as agriculturists rather than as master and servant, the satisfactory completion of the most important task in the farming calendar. The Harvest Festival was no substitute, for it was essentially a public occasion when the social status of the participants was emphasised not held in abeyance. The labourers were entertained as a class, the 'lower orders', not as individuals in a private gathering. The harvest home was essentially a time for relaxation from the cares of work, for the long weeks of continuous labour were over, and from social conventions, for everyone was licensed to drink their fill. For one evening at least the need to 'hold tongue' was absent and if drunken words could carry a sting, they would bear no consequences.

But the tone of reports of harvest homes as the century advanced almost suggests that they had become anachronistic in the circumstances of a wage economy and a rigidly structured hierarchical society. The fact that on certain farms they were being revived when they were still a continuing though declining custom suggests a self-consciousness that altered the whole content of the custom. The impact of the Harvest Festival on the actual form of the celebration suggests a change in the purpose of the custom. The days of identification of labourer and farmer in a common task were already over.

NOTES

1 Cf. William Coles Finch, 'Life in Rural England' (1928), p. 56; Flora Thompson, 'Larkrise to Candleford' (1962 edn), p. 14.
2 The English Reports, vol. CXXVI, Common Pleas IV, p. 32, H. BL 51.
3 Robert Gibbs, 'A Record of Local Occurrences', 4 vols. vol. III (1882), p. 133; C. S. Orwin and E. H. Whetham, 'History of British Agriculture 1846-1914' (1964), p. 175, 'it was not until 1883 that tenants were legally allowed to shoot hares and rabbits on their own farms'.
4 English Reports, p. 32.
5 Thompson, 'Larkrise to Candleford', p. 14. Leasing is the older word for gleaning; Bod. MS Top Oxon. C.515 L.N. Letch Bucknell Village, 1933 (no. of additions 1935-6), f.33, 'In Harvest time the people go leasing'; Bod. MS Top Oxon. d.475 History of Filkins (G. Swinford 1958), f.4; 'The miller kept bran and toppings to pay for grinding'.
6 Mabel K. Ashby, 'Joseph Ashby of Tysoe' (1961), p. 25.
7 J. C. Cornish, 'Reminiscences of a Country Life' (1939), p. 7.
8 Walter Rose, 'Good Neighbours', (1942), p. 30.
9 English Reports, p. 37, 1 H.BL 60.
10 Ibid., pp. 37-8, 1 H.BL 60-62 (J. Heath).
11 Ibid., p. 38, 1 H.BL 62 (J. Wilson).
12 See Mrs Bryan Stapleton, 'Three Oxfordshire Parishes' (Oxford, 1893), p. 167.
13 OJ, 7 Oct. 1809.
14 'Bicester Herald', 26 Aug. 1887.
15 Gertrude Jekyll, 'Old West Surrey' (1904), p. 188. 'Gleaning... allowed before the sheaves were carried for those whose fathers were harvesting; after the corn was carried anybody was usually allowed to glean'; N & Q, 4th Ser. (Sept. 1869) - similarly about Alford Lincs; but only 'own' men in any field of 'cleared corn' in Notts; in E. Lincs unrestricted.
16 'Aylesbury News', 29 Aug. 1840.
17 But cf. A. J. Spencer, 'Dixon's Law of the Farm', 5th edn (1892), 'it is still illegal within 5 Rich 11 St. 1.C.8. to turn out the trespasser with violence' (Edwick v Hawkes 18Ch D 199); also 'damages may be recovered even by a person whose possession was wrongful' (Beddall v Maitland 17Ch D 174).
18 JOJ, 17 Sept. 1859.
19 Ibid., 4 Oct. 1862; cf. ibid., 27 Sept. 1862, Wickham woman leasing barley when forbidden - case dismissed (Banbury Division Petty Sessions, 25 Sept).
20 OT, 11 Oct. 1890.
21 Finch, 'Life in Rural England', p. 56: gleaning a privilege granted to regular workers (Kent); Rose 'Good Neighbours', p. 30: usually wheat only.

22 N & Q, 2nd Ser., vol. X (15 Dec. 1860), p. 476.
23 Ibid. (3 Nov. 1860), p. 358.
24 Ibid., 9th Ser., vol. VII (7 Sept. 1901), p. 268.
25 A. H. Cocks, 'The Church Bells of Buckinghamshire' (1897), p. 279.
26 William Andrews, 'England in Days of Old' (1897), p. 158. 'At Lyddington, Rutland... the [parish] clerk claims a fee of a penny a week from women and big children; at West Deeping, Lincolnshire... two pence a head... as they refused to pay... declined to ring the bell.'
27 'Folk-lore', vol. XXII (1911), p. 237.
28 Cocks, 'The Church Bells of Buckinghamshire', pp. 279, 303, 392, 543, 556, 564.
29 Winifred H. Beaumont, 'The Wormingford Story' (Wormingford, 1959), p. 22; so called in Essex; cf. George Edwards, 'From Crow Scaring to Westminster', p. 25; 'at a given hour the farmer would open the gate... remove the sheaf and shout "All on" ' (East Anglia).
30 Personal Communication, GAR 1, Walter King, Garsington.
31 N & Q, 2nd Ser., vol. X (13 Oct. 1860), p. 285; but in contrast 'Reading Mercury', 16 Sept. 1893; 9-year-old girl gleaning barley assaulted by jobbing labourer from Ashampstead in a field at Aldworth.
32 Cocks, 'The Church Bells of Buckinghamshire', p. 379.
33 Agnes Gibbons and E. C. Davey, 'Wantage Past and Present' (1901), p. 94.
34 Alfred Gatty, 'The Bell' (1848), pp. 18-20.
35 Cocks, 'The Church Bells of Buckinghamshire', p. 537, also at 1 p.m. (dinner bell) and 8 p.m. (curfew bell) every weekday throughout the year.
36 Cf. Gatty, 'The Bell', p. 42 - the curfew bell was rung at Carfax, Oxford; Gibbons and Davey, 'Wantage Past and Present', p. 94 - curfew still rung at Wantage, 1900.
37 'Customal (1391) and Bye-Laws (1386-1540) of the Manor of Islip', Barbara F. Harvey (ed.) (Oxfordshire Record Society, 1959), p. 109.
38 'Warwickshire Advertizer', 29 Aug. 1848. Extract from letter to 'The Times', Saturday 12 Aug. 1848, signed S. G. Osborne.
39 N & Q, 2nd Ser., vol. X (Dec. 1860), p. 519: prohibition on gleaning before 8 a.m. announced from pulpit at Gillingham, Norfolk, c. 1855 - otherwise 'people would go into the fields too soon'; N & Q, 2nd Ser., vol. X, no. 1 (26 Jan. 1861), p. 78: 'The want of a gleaners' bell was so much felt that the 9 o'c Mass bell of the Catholic Chapel has been long adopted as the signal for the gleaners to start fair together'; Henry Best, 'Rural Economy in Yorkshire in 1641' (Surtees Society Publication, 1857), p. 46: 'errour... to rake winter corn... a few of those raking will... blacken and spoile a greate deale of better corne [for] grownde... always dirty dusty and foule'.
40 Alfred Williams, 'Round About the Upper Thames' (1922), p. 142.

41 But see Thompson, 'Larkrise to Candleford', p. 259: '[a tradesman] used to say the farmer paid starvation wages all the year and thought he made it up by giving that one good meal'.
42 Angelina Parker, Oxfordshire Village Folklore 1840-1900, 'Folk-lore', vol. XXIV (1913), pp. 85-6.
43 Williams, 'Round About the Upper Thames', p. 143.
44 Parker, Oxfordshire Village Folklore, p. 82
45 J. S. Udal, A Dorsetshire Harvest-Home, N & Q, 4th Ser., vol. XII (20 Dec. 1873), p. 492.
46 Clare College MS C. Sharp Folk Dance Notes, vol. II, f.94.
47 Parker, Oxfordshire Village Folklore, p. 83.
48 E. E. Street, Sussex Harvest Home, N & Q, 4th Ser., (19 Oct. 1872), p. 312.
49 Bod. MS Oxon C22. 'The Harvester Songster being a collection of Choice Songs to be sung at Harvest Home', printed and sold by J. Pitts, Seven Dials, price one penny (no date but c. 1820).
50 Bod. MS 1472, f.21. 'Special Harvest Hymns' (Geo. Burroughs, 1901); 'Come ye thankful people come' was in fact sung at one of the earliest harvest services held in Archdeacon Denison's parish in 1862 - N & Q, 3rd Ser., vol. II (15 November 1862).
51 Clara J. Jewitt, 'Folk-lore', vol. XIII (1902), p. 180.
52 Parker, Oxfordshire Village Folklore, p. 85.
53 B. Lowsley, 'A Glossary of Berkshire Words' (1888), p. 14; cf. Andrews, 'England in Days of Old', p. 251, but 'Weel bun and bettersshorn, is Master ——'s corn'... shouted by Cleveland labourers forking the last sheaf on the wagon' (p. 248).
54 Consule Planco, The Last Load, N & Q, 4th Ser., vol. X (2 Nov. 1872), p. 359.
55 Cuthbert Bede, The Last Load and Harvest Home, N & Q, 4th Ser., vol. X(12 Oct. 1872), pp. 286-7.
56 N & Q, 4th Ser., vol. IV (25 Sept. 1869), p. 253.
57 Andrews, 'England in Days of Old', pp. 249-51; but cf. Richard Blakeborough, 'Character, Folklore and Customs of the North Riding of Yorkshire' (1898), pp. 85-6. Note, when the last sheaf was made into a decorated effigy and used ceremoniously it was called 't'mell doll', otherwise it was sometimes called the 'widow' (Carthorpe), both terms extant in 1898 as well as the 'mell' (harvest) supper.
58 N & Q, 6th Ser., vol. IV (1881), p. 127; ibid., p. 186; ibid., 5th Ser., vol. VI (1876), pp. 286, 336; ibid., vol IX (1878), p. 306; ibid., vol. X (1879), p. 359; ibid., 9th Ser., vol. I (1898), p. 303; Andrews, 'England in Days of Old', p. 248; Alice B. Gomme, Notes on Berwickshire Harvest Customs, 'Folk-lore', vol. XIII (1902), pp. 185-6; ibid., vol. XXV (1914), pp. 379-80.
59 N & Q, 6th Ser., vol. IV (1881), p. 127.
60 Andrews, 'England in Days of Old', p. 250.

61 N & Q, 6th Ser., vol. IV (1881), p. 186; ibid., 5th Ser., vol. VI (1875), p. 286 and 336; ibid., vol. IX (1878), p. 306; ibid., vol. X (1879), p. 359.
62 Ibid., (1898), pp. 303-4. Note 'frog' considered another form of 'frock' and so equivalent to 'neck' (corruption 'knack'), both 'frock' and 'neck' implying 'plaiting' - in the old smock-frock the 'frocking' was the plaited ornamentations. 'Cutting' the neck or mare or other terms like cock or hare, evoked by the semblance produced by plaiting, and likewise 'Crying the Neck' etc. were recorded customs - no mention of 'Crying the Frog'. 'The 'cutting' clearly was one thing, the 'crying' was another' (p. 304).
63 Alice B. Gomme, 'Notes on Berwickshire Harvest Customs', p. 178.
64 J. G. Frazer, 'Spirits of the Corn and of the Wild' (The Golden Bough, vol. III, 1912), see Ch. VIII, The Corn Spirit as an Animal, pp. 270-305; cf. N & Q, 8th Ser., vol. IX (1896), p. 128, in Lincolnshire c. 1825 the 'old sow' - two men dressed up in sacks visited the harvest supper and pestered the guests - its head filled with furze cuttings. 'She related to Gullimburstii, the boar which drew the car of Fury, in Norse mythology.' G. is said to typify the fields of ripe corn, over which Frey is lord as bestower of sunshine and rain and protector of crops. See also N & Q 8th Ser., vol. IX (1896), p. 176 - monster 'Paiky' in Dumfriesshire. 'The Golden Bough' (Abr. Edn 1967), pp. 449-50: 'the custom of throwing water on the last corn cut... or on the person who brings it home... observed in Germany and France, and till lately in England and Scotland... avowed intent to procure rain for the next year's crop'.
65 C. W. von Sydow, Mannhardtian Theories about the Last Sheaf and the Fertility Demons, 'Folk-lore', vol. XLV (1934), pp. 291-309; cf W. Mannhardt, 'Die Korndamonen' (Berlin, 1868) and 'Kind und Korn Mythologische Forschungen' (Strasburg, 1884); see N & Q (1898), p. 304: 'it is plain in the past there was great pride taken... in being the first to accomplish any agricultural work... many can recall the loud and prolonged cheering... raised by the work people... who were first to finish harvest'.
66 Clare College, Cambridge MS C. Sharpe, Folk Dance Notes, vol. 1, f.55. An account of 'Crying the Neck' given by F. Hancock of Minehead, Somerset c. 1895.
67 Ronald Blythe, 'Akenfield' (1969), p. 56. 'The Lord sat on top of the last load... Master would... kill a couple of sheep for the Horkey supper and afterwards we all went... [S]houting in the empty old fields... the boys in the next village would shout back' (John Grout, aged 88).
68 'Ag. Gaz.', 30 Apr. 1860, p. 392. Devon: ordinary rate 11s per week, harvest rate 11s per week with meal and drink; Shropshire: 10s per week and dinner; Suffolk generally £5 in cash, 10 pecks of malt, 1 stone of pork. Such payments excluded any piece work.

69 J. C. Atkinson, 'Forty Years in a Moorland Parish' (1891), p. 241, ' "kern" supper, "kern" feast, "kern" baby, referred to the word corn... "mell" in "Mell" supper, "mell" sheaf... meal, a service of food, mêler (Fr.) to mingle, melee a contest'.
70 N & Q, 6th Ser., vol. III (1880), p. 469. Largesse asked for at Cromer, Norfolk on several occasions either by reapers or children. Note: still 'called for in a loud voice without loss of dignity' c. 1930, see H. Barrett, 'Early to Rise' (1967), p. 125.
71 N & Q, 6th Ser., vol. IV (1881), p. 193; ibid., p. 259. 'Robert Bloomfield [Poems, 1809] explains Horkey as harvest festival... an account of "the lord"... The "frolic" or "largesse-spending"... the shout "largesse"... customs fast going out of use. But it would seem they are not yet quite done.'
72 William Hone, 'The Everyday Book', vol. II (1827), p. 1183.
73 Hone, 'The Everyday Book', p. 1154.
74 Ibid., p. 1063. 'Blound the glossographer says that Lammas is called Half-Mass that is Loof-Mass or Bread Mass which signifies a feast of thanksgiving for the first fruits of the corn - New Wheat called Lammas-wheat'.
75 Ibid., p. 1059, quoting from Leigh Hunt's 'Months'.
76 E. Chambers FRS, 'Encyclopaedia' (1779), vol. II.
77 Lawrence Whistler, 'The English Festivals' (1947), p. 182. 'Nothing like the modern Harvest Thanksgiving seems to have existed in Mediaeval times.'
78 Richard Jefferies, 'The Toilers of the Field' (1892), p. 262. 'Harvest is a time of freedom, but the last day resembles the ancient Saturnalia, or rather perhaps the vine season in Italy when the grape gatherers indulged their rude wit on everyone who came near.'
79 Whistler, 'The English Festivals', p. 193.
80 S. Baring Gould, 'The Vicar of Morwenstow' (1876), p. 226.
81 Whistler, 'The English Festivals', p. 182.
82 Cf. John R. H. Merrman, 'A History of the Church of England' (1953), p. 380, 'an eccentric Cornish priest R. S. Hawker... introduced the custom...'.
83 N & Q, 7th Ser. (27 Nov. 1886), p. 425.
84 Bod. MS 1419 f.1316 (33), The Harvest Festival at Lilbrook (1861), p. 2.
85 N & Q (27 Nov. 1886), p. 425.
86 JOJ, 23 Sept. 1865.
87 Gibbs, 'A Record of Local Occurrences', vol. IV (Aylesbury), p. 161.
88 'Bucks Herald', 27 Sept. 1873.
89 'Illustrated London News', 10 Oct. 1863, p. 369.
90 Bod. MS, The Harvest Festival at Lilbrook, p. 17.
91 JOJ, 1 Oct. 1859. 'Ashendon man and wife charged with disturbing the peace... at Brill Magistrates Meeting... they had been to a Harvest Home...'

92 Until Ackland's Agricultural Wages Bill of 1887 payment in beer was forbidden, but still practised.
93 Cf. Dunkin's 'History of Bicester' (1816), pp. 269-70.
94 Arthur Young, 'The Farmer's Calendar', 6th edn (1805), p. 421. Earnest - the 1s sealing a 'harvest contract'. But it varied, see Reading Farm Records, BER 13/5/5, Babcock Wick Farm, Radley, Labour Book, 19 August, 2s also; BER 28/3/1, Bradley Hill Farm, Chieveley, Labour Book, 18 labourers, 2s 6d each, 8 women 1s 6d, 6 lads 1s 6d; CAM/1/1/1, Underwood Hall and Partridge Farms, 1891 and 1893, 2s 6d each harvester is to equal a horkey.
95 Bod. MS Top Oxon d. 199 Manning Papers, f.368.
96 Oxford RO, DIL 1/q/9c, Ditchley Estate Labour Book, November 1876 (Fulwell Farm), entry for 27 Sept. 1878 also 1s for women; see also JOJ, 12 Oct. 1872, Ditchley Estate Harvest Home about 50 sat down to a dinner of the 'Good Old English type'.
97 See p. 113.
98 JOJ, 4 Oct. 1862.
99 See N & Q, 6th Ser., vol. IV (3 Sept. 1881), p. 193.
100 Bod. MS Top Oxon d 199, p. 368.
101 'Reading Mercury', 26 Aug. 1893. Sir Walter Gilbey appealing for funds explained, 'persons who have cultivated holdings varying from 100 to 2,000 acres [are currently] seeking election to the benefits of the... Institution'.
102 N & Q, 9th Ser. (24 Mar. 1900). Report of death of Rev. John Soing - correspondent of the 'Church Times' - first clergyman to introduce harvest festivals in London ('Globe' of 13 Jan. 1900).
103 JOJ, 21 Sept. 1872.
104 Ibid., 5 Sept. 1872.
105 'Reading Mercury', 14 Oct. 1882.
106 Ibid., 21 Oct. 1882.
107 'Bucks Herald', 12 Oct. 1889.
108 'Reading Mercury', 30 Sept. 1893.
109 JOJ, 1 Oct. 1898.
110 OT, 4 Oct. 1890.
111 'Reading Mercury', 16 Sept. 1882.
112 Ibid., 23 Sept. 1893.
113 Bod. MS Top Oxon d 199, p. 368.
114 'Reading Mercury', 30 Sept. 1893.
115 OT, 5 Oct. 1893.
116 Ibid., 24 Sept. 1889.
117 Ibid., 28 Sept. 1889.
118 JOJ., 17 Sept. 1898.
119 Ibid., 23 Sept. 1899.
120 Ibid., 30 Sept. 1899.
121 J. Arthur Gibbs, 'A Cotswold Village' (1898), p. 387; Andrews, 'England in the Days of Old', pp. 253-4.
122 N & Q, 6th Ser., vol. IV (1881), p. 492.
123 'Labourers' Union Chronicle', 18 Oct. 1893.

10

CONCLUSION

> Inequality is not merely a matter of individual abilities and aptitudes, it is above all a social fact. The opportunities an individual has and even his abilities are in part governed by his position in society. While many have talked about classless societies of either the past or the future, their ideas have found very little support in historical experience.
>
> André Béteille, 'Social Inequality' (1969), p. 15.

During the nineteenth century the fate of the agricultural workers in the southern half of England was to be doubly sealed. Skilled or unskilled, prepared to work on some large estate or farm, offering subservience in exchange for some security and an indeterminate economic gain from a declining paternalism, or risking casual employment for a dubious freedom, they were to become fixed in low-paid employment.[1] The industrial expansion of Great Britain depended on a policy of cheap food for a growing urban population. Despite migration to the centres of industrial development, despite emigration to the empty spaces of the New World, there was still a growing rural population. Consequently those who remained were not in a position to bargain for their labour - only to accept work at a rate dictated by the law of supply and demand.

Economic forces decreed that this wage should be at or near a subsistence level.[2] In addition, social and religious attitudes reinforced the maintenance of a class structure - for social stratification was essential to social stability as well as to economic growth.[3] A reviewer of 'Joseph Arch's Book' reflected that:

> [T]he Church minister of that day [1872] did not... take sides in any social upheaval; he waited to see; and was by association and training against any agitation whereby the existing state of affairs would be disturbed.[4]

That it was unthinkable was due to the hold of the classical economists on the social thinking of the period. Laissez-faire capitalism held out not so much the prospect of success but the threat of failure.

> Progression in the age of Malthus, Ricardo Senior and the two Mills was seen either as impending deterioration or else as a movement towards a stationary state.[5]

'It was no wonder', suggests C. B. Raven, 'that the first unpleasant reminder of their animal ancestry was forgiven [Darwin].'

> The well-to-do classes had already accepted the survival of the fittest as their own justification. To discover that it was the universal law of nature responsible for all progress and operating with ruthlessness for the ultimate good of the race, was to find their prosperity given a halo of sanctity...[6]

The fact that the rural wage therefore became inextricably tied at a level below the level of the industrial worker burdened the conscience of very few.[7] It is understandable that Lady Bowley in Dickens's story 'The Chimes' (1846) could get her villagers singing:

> O let us love our occupations,
> Bless the Squire and his Relations,
> Live upon our daily rations,
> And always know our proper stations.[8]

In the early decades of the century the rural workers, deprived of common rights by the enclosure acts[9] and deprived of alternative employment through the decline in home industries, were finally transformed into wage earners in a seasonal occupation, which itself had changed from the remnants of a system of subsistence farming into a full-scale commercial enterprise. An added irony of those years was the operation of the Speenhamland system of poor relief which instead of elevating the position of the 'deserving' poor in a changing society provided them with two masters who between them destroyed their dignity and the value of their work.[10]

In addition, in the final years of the century, the marginal gains which an emergent trade unionism in the early 1870s had procured for them against fierce opposition and a background of general rising prosperity were lost, while any coercive power left to the agricultural labourers from the economic strength which a declining pool of labour would previously have given them was pre-empted by the threat of an increased use of machinery, a threat indeed already implemented when strike action for better conditions had been resorted to.[11]

It is true that without resources other than a wage both the industrial and the agricultural worker in the nineteenth century were faced with a similar problem in the fluctuating labour market. The cycle of expansion and recession in trade loosely fixed the opportunities of industrial workers for finding work and of bargaining a more favourable wage rate. The cycle in agriculture however was contained in the seasonal winter dearth and certainty of work at the busy season in summer. With a growth of rural population this cycle became more tightly containing in winter, more explosive at harvest time, because the achievements of the Agricultural Revolution were essentially related to methods of cultivation and farm reorganisation leading to increased production. The increased acreage of corn crops was more vulnerable to adverse weather conditions, there was a greater risk of simultaneous ripening; carting from more distant fields to the safety of the stackyard took longer.[12] These contingencies had to be insured against in the calculations of a farmer's labour requirements and often meant immediate engagement of casual labour if it could be found. But more importantly there was still no certainty that all available resources would be sufficient, for mechanical aids were minimal, and the tools available in the first half of the century were the hand tools of a peasant economy. They were neither factory-made nor designed to cope with the dimensions of a 'manufactory's' production. Each year the ripening corn grew into profit; only a vast army of harvesters could make that profit secure.

The harvest then provided work for all; it restored family earning power; it altered the circumstances of competition - no longer necessarily worker against worker for limited work - thus giving the labourers the only opportunity to bargain their skills and restoring dignity to them in the use of those skills. The magnitude of the harvest and the uncertainty of its gathering created a new dimension of precarious freedom in their lives. Each summer they looked to the extra harvest earnings to provide for more than the payment of past debts and the means to tide them over the hardships of winter. If they believed in the myth of better days or sought consolation that they were active participants in an 'holy time',[13] only when enough of them realised that it was a myth, and departed, could the myth grow into something approaching a reality.

Faced with a new set of practical problems, arable farmers by the 1850s were provided with an obvious solution - the early adoption of mechanical reaping - had harvesting been solely a technical operation in one of the stages of corn production. But the development of harvesting involved a complex interaction between social, economic and traditional forces within the distinct circumstances of society which were neither necessarily ubiquitous nor even similar in the adjacent counties of rural England. These circumstances initially controlled the pattern of change, but in turn were themselves changed. In the extreme case of the expansion of the corn-growing lands of America the scanty population made the rapid adoption of machinery essential. In the north of England its adoption followed an early absolute shortage of hand labour. In southern England the situation was entirely different. A still-growing rural population, inhabiting the open villages created by the policy of the farming community not to maintain cottage accommodation except to house the often minimal numbers of annually hired labourers, produced a pool of surplus labour to meet harvest requirements.[14] For example George Culley described Wootton in the 1860s as:

> a large open parish with many very poor cottages, both in the village of Wootton and in Old Woodstock. A very small proportion of the cottages belong to the landowners. The largest estate has seven cottages for 1,570 acres let in farms; the rector has 10 for about 700 acres, another proprietor has two for 540 acres; and another has 400 acres without a cottage. Balliol College has one cottage for 320 acres; the churchwardens have 15 cottages... Rents vary from 52s to 80s.

The parish at the time contained 1,238 inhabitants with 3,270 acres, chiefly in arable cultivation.[15] With regard to Tackley with a population of 626 and the parish containing 2,850 acres under mixed cultivation, the Rev. L. A. Sharp at this period reported that the most important resident had built some excellent cottages, even if the rents were rather high. Culley however found here:

> One estate having about 600 acres in farms has 26 cottages
> One estate having about 600 acres in farms has no cottages
> One estate having about 500 acres in farms has 15 cottages
> One estate having about 400 acres in farms has no cottages
> Christ Church College has 350 acres with no cottage
> Cottage rents varying from 40s to 80s.[16]

Of the cottages at Cassington with a smaller population of 433 but with a similar acreage to Tackley under mixed cultivation (2,990 acres), the description given by its vicar was that they

> vary; some are tumbledown and little better than piggeries; landlords of such an abomination ought to be subject to severe fine. The rent exceeds by far the value of such tenements.

Culley added the following information:

> [T]he chief landowner in this parish has about 20 cottages for 1,140 acres of farm land (eight or nine of the best of which are occupied by railway labourers). Another estate of 300 acres has five cottages; the remaining cottages belong to small tenement holders. Ch. Ch. Coll. Oxford has 80 acres without a cottage. Cottage rents vary from £2 to £5 a year.[17]

The main body of farmers therefore not only had the advantage of a low wage market, created by the seasonal demands of arable farming, but avoided contributing, to any extent, to its maintenance, for upon the open parishes fell the burden of the poor rates which supported the underemployed and unemployed during the winter months. In many cases, therefore, wages, low as they were, did not reflect the real cheapness of harvest labour. Technological change in the use of hand tools, so that full advantage could be taken of this cheap labour, rather than mechanisation involving capital expenditure was therefore an economically sound policy. The social consequences of such a policy were an important consideration for the amount of harvest work controlled not the extent of the labourers' affluence but the degree of his poverty.

That this is no exaggeration is evident from the tightly constrained economic position revealed by their budgets; their most basic needs virtually absorbed the total of their cash resources. A day labourer of Harbury, Wawickshire, was considered in 1873 sufficiently well off to contribute 1s per week towards the support of his indigent parents. He had been fortunate enough to have had regular employment at 15s per week; with extra earnings at harvest time, profit from his allotment put at £2 18s and the value of vegetables at £2 10s, his income was estimated at £46. A breakdown of his budget, however, revealed a deficit of £7 11s 6d.

	s	d
Rent	2	0
Coal	1	6
Bread	3	1½
Grocery	3	0
1 Gallon of Flour		7
Cost of 2 pigs		1½ [spread over the year]
Feeding same	2	6
Club money		6
For wife and child		1½
Man's clothes	1	6
Wife & child's clothing and bedding	1	0
Butter 1 lb	1	4
Cheese ½ lb		5
Beer for Sunday supper		2½

<u>Total</u>: £50.14s[18]

A Sussex family in the 1870s consisting of day labourer, wife and six children had a similar weekly income (15s 6d) to subsist on. Because of the wife's reputaton for frugal management the vicar of Burwash of that day, concerned with rural poverty, obtained these details of the family's expenditure:

	s	d
Rent	2	0
7 Gallons of Flour	7	0
2 lb Dutch Cheese	1	3
1 lb Butter	1	4
½ lb soap		2
Soda ½d Blue ½d		1

	s	d
Salt and Pepper		½
1½ lb candles		10½
2 oz. tea		4
2 lb sugar		7
Schooling		7
Cotton 1d Mustard etc. 2d		4
Milk per week - viz. ¼ pint of skim daily		3½
Mangling		1
	14	10½

It is to be noted that almost half the income only covered flour, and would have been more if spent directly on bread - meat is not mentioned, or fresh milk, or beer or tobacco. 'Outgoings for wood, clothing, boots', however, came from the 'extra earnings by piece work... cutting... mowing... harvesting.' Any surplus from these earnings together with the 7½d over from the basic weekly expenditure could have remedied this budget's obvious inadequacy. By any standards it was frugal.[19] In all such budgets the importance of harvest earnings is emphasised while the proportion of income spent on bread or flour highlights the difference that gleaning could make in releasing cash for other needs.[20]

If in later years in the century allotments proved a great boon, their continuing disadvantage was that they needed some initial cash outlay, while, when they became more widespread, the value of the produce though providing food for the household depreciated in cash terms. An Ilmington labourer whose father had worked as a quarryman had this to say in the 1890s:

> Employment was not good so our allotment had to suffice. It provided plenty of food but no money for groceries, clothing or rent. Dad's quarry had closed down, as by this time stone could be bought already broken. Though he was a skilled farm worker, hedger, rick-builder and thatcher, the farmer's pleaded bad prices for offering no work.[21]

While it is tempting to apply the orthodox economic principle of supply and demand to account for a continuing low level of wages and the extent of poverty in arable areas, it would mask the essential factor that agricultural labour at harvest time in the nineteenth century cannot be identified as a fixed group of workers. In winter there was a clear-cut division between hired labourers, whose annual wage was fixed at Old Michaelmas when the supply of labour was elastic, and the day labourers whose numbers ensured that this rate was low. In summer, therefore, the extra work performed by hired labourers was paid for at a rate linked to an already fixed low rate while the remaining labour force was not simply the day labourers, but family labour, migrant and casual workers and even workers from other trades. Thus an absolute inelasticity in the labour market with consequent pressure on wages which normally would have been present with a fixed labour force did not occur. If, however, a considerable proportion of the extra summer earnings accrued to the day labourers through individual effort and family participation, there could be little economic advancement, for these earnings, for the most part, served to balance the periods of winter underemployment or unemployment. Similarly, with no effective means of obtaining higher annual wages and with more limited earning opportunities at harvest time, hired labourers were also prevented from individual advancement. In either case a continuing poverty was unavoidable.

The day labourer was a creation of the early years of the nineteenth century; upon his labour and the continuing peasant practice of family participation depended the successful gathering of the century's harvests. As it drew to a close, whenever and wherever the reaper-binder took over the early stages of harvesting which had previously provided the day labourers's most lucrative work, both family earnings and the presence of the day labourer declined. It was in the 1890s[22] that the erstwhile independent labourers or their sons finally deserted the countryside, as the pattern of twentieth-century farming based on hired labour, harvest machinery and the tied cottage began to emerge. Independence was no longer precarious; it was becoming almost impossible if reliance on harvest earnings was looked to for survival. It was in the 1890s that the local newspapers began to carry more and more advertisements for hired labour,[23] that the village schoolmistress missed familiar faces in the schoolroom and had to, almost for the first time, enter strange names in the school register.[24] By the turn of the century the independent labourer ceased to be in the majority. Those that did not depart had increasingly to find a hiring, but there is a new note of resentment in their recorded utterances[25] which denotes that the loss of an independence however precarious was bitterly felt:

> The farmers were sharp with us. If you couldn't do a job you were reminded that plenty more could. So you had to be careful. I had to accept everything my governor said to me. I learnt never to answer a word. I durs'nt say nothing. Today you can be a man with men, but not then... I lived when other men could do what they liked with me. We feared so much. We even feared the weather!... we were always being sent home. We dreaded the rain; it washed our few shillings away.[26]

The nineteenth century saw farming become a commercial enterprise instead of a social activity. The new owners of property, with their wealth accumulated from industrial enterprises, saw land as an investment for profit and not as a trust held for the welfare of others. The old landed gentry forsook paternalism in an effort to compete. The worker suffered in the transition for while deprived of the benefits of 'charity' he found it impossible to obtain any commensurate gain in dignity or in material advancement through his own efforts. The hierarchy of the countryside still proclaimed the virtue of work and of thrifty living, when the kind of work provided was lengthy toil with little reward, and when thrift offered no expectation of eventual 'embourgeoisement'.

Before enclosure the work was there, and the common land was there to translate work into reward. Some reward accrued to not just part of the community but to each striving individual, and the gain was from the land, not from the exploitation of the labour of others. Now the reward had to come from the employer, but the right to something more than a subsistence was continuously denied. Now even the opportunity to work was in other hands. Historically, the enclosure of land was inevitable. The progressive ideas and practices, which fashioned what is loosely termed the Agricultural Revolution in the eighteenth century, could not have effected any change in nineteenth-century agricultural practice within the old framework of strip-tillage, and unsystematic, unscientific and wasteful use of common lands.[27] Though cheap labour and bad working conditions were not economic, the farming fraternity feared that if the working man became more prosperous he would no longer be content to labour long monotonous hours in the field. And there was no compulsion to debate the need or desirability of Bentham's 'Panopticon' since the poor were already occupying the role of those prisoners which his projected model prison would have held. The landowner, the squire, the vicar and the tenant farmer perpetually enjoined principles, but they most assiduously proclaimed those that redounded to their own advantage. They gathered to themselves the wealth from the years of good harvests and extended little help to those who suffered most from the years of the bad ones.

The idealisation of the English countryside did not appear in English literature until men could look in upon the farm and the field from the outside to perceive its pleasantness, and ignore the grievous realities. Many Victorian and later writers of the countryside[28] wrote from one end of the social spectrum, seeing the squire and the high farmer as the fathers of their people, sentimentalising ephemeral social harmonies, and ignoring all the inhumanity and exploitation.

Marianne Farmingham wrote a poem in 1900 entitled 'The Corn'. These are the last three stanzas:

> It has grown on in its quiet stateliness,
> Blessed of the heart, unhelped by hands.
> Until, with beneficient golden plenty
> It crowns and covers our English land.
>
> Now for the sound of the merry sickle
> In the cool grey morning before men roam.
> And the gladder sound of the grateful singers
> Joining the chorus of Harvest Home.

> Now for the folding of hands in thankfulness
> God has heeded the prayer we said;
> Happy are we for the gracious Father
> Will each day give us our daily bread.[29]

All the evidence suggests that harvesting in the main corn-growing areas was the one major factor that influenced social and economic attitudes and change in nineteenth-century southern England. In other areas there was some possibility of economic and social mobility, but not so in these rural communities. While enabling a growing number of workers to remain attached to the soil, the checks and balances within and surrounding harvesting for profit within an agricultural situation threatened by industrial inroads on labour, and a continuing low wage economy, perpetuated for socio-economic reasons, provided little other social or economic advancement. Durkheim in 'The Division of Labour in Society' had this comment to make on rural societies in general:

> As long as industry is exclusively agricultural, it has in the family and in the village... its immediate organisation... is sufficiently regulated by the family, and the family itself thus serves as an occupational group.[30]

No greater contrast can be found than in the southern counties in this period when winter work opportunities were so totally in the hands of commercial farmers. There was not so much a minimal, as an optional demand for labour in the winter months, depending solely on the individual farmer's consideration of what was good farming practice, his assessment of what expenditure on labour, on cultivation, on use of fertilisers was needed, or was economic in terms of outlay in relation to anticipated harvest returns. The amount of work that hedging, ditching and drainage provided was a farmer's decision.[31]

Neglect cut immediate work opportunity but farming profits were jeopardised only in the longer term. A further curtailment resulted from the fact that many day-to-day farm operations, though increasing, could be adequately performed by the cheaper labour of women and juveniles, and the work itself - in the stables, leading the team at plough at planting time, stone picking, weeding and hoeing - was often considered by farmers mainly as other than men's work. Thus the number of able-bodied paupers in receipt of poor relief reached its peak in February, and only showed a marked decrease during the harvest months.[32]

Without an individual improvement in economic position, there was little hope of an improvement in social status within Victorian Society.[33] But if in the tight fabric of village life, work and charity, there were a hundred small ways of putting pressure on those who forgot their proper station, there were ostensibly two specific checks on economic progress - the impact of the settlement clauses of the poor law and the law relating to hiring agreements. On the face of it, a less rigid approach to removing the unsettled seeking poor relief developed progressively from the introduction of the New Poor Law (1834) and encouraged a less obsessive regard by labourers for the maintenance of a settlement which previously had narrowed down both choice of work and the incentive to seek work further afield.[34] Yet the restriction on mobility may have subsequently been largely inconsequential within the internal rural situation. How far a change of abode within the arable counties could have improved a labourer's economic position is not clear; winter work opportunities were little different from one village or for that matter one county to another in any permanent sense. Alternatively, in summer, widespread temporary harvest migration posed no settlement problem; indeed it pre-empted the traditional purpose of settlement - the retention within village communities of adequate harvest labour.[35]

The increased severity of the penal code relating to breach of contract containing the threat of three months imprisonment, previously one month, dating from 1824 (4 Geo. IV c.34), continuing implicitly until 1875 (38 & 39 Vic. c.90) and explicitly, within the Conspiracy and Protection of Property Act, 1875, to the present day had more compulsive impact that the poor law. While it reinforced the farmers' control over both annually hired and casual harvest labour, it reduced any incentive skilled hired labour might have had to obtain the real value of their labour at harvest time when the market was for once in their favour. Even so the breaking of agreements whatever the penalty, the rejection of paternalism, though economically advantageous, because it implied an acceptance of social inferiority, the eruption of trade unionism continuing into the high summer of the early 1870s, and which finally exposed the process of harvesting as no more than and no different from a part of any commercial undertaking, are evidence enough that some independence, some recognition of social status was not only valued, not only demanded, but possible. Without, however, the distinct and separate organisation of work, and the special opportunities for cash earnings at harvest time, this show of independence could not have occurred.

If harvest customs lingered longer in the northern counties of England sustaining individuals with a continuing spirit of community, piece work in the south provided some economic recompense. If these harvest earnings depended on excessive individual and family effort, the amount of work and the skill with which it was performed brought recognition and esteem from local inhabitants. If a tendency to excessive drinking depleted these earnings, nevertheless it reinforced community ties. And if access to land in the shape of allotments only began to increase in the last decade of the century, farmers whatever their wishes or their legal rights failed in the main to prevent the annual invasion of their fields which the continuing custom of gleaning produced. Nineteenth-century harvesting underlined a 'culture of poverty' and this twentieth-century sociological concept is brought alive if only by those applications and expressions and recorded customs which in the first decades of a new century ceased to be of moment, but previously were a major part of rural existence. But the 'Lord of the Harvest' and 'gleaning bell', such terms as 'task' work, 'widow' man and 'earnest', such exclamations as 'It was like magic', the 'rites' surrounding the 'Last Sheaf' and the abandon of the 'harvest home' were as anachronistic in a wage economy as were 'payment in kind', the Settlement laws and the Labour laws in a country committed to the free movement of the market. Yet so long as unfavourable economic circumstances for the agricultural labourer persisted, so long were needed these cultural compensations. It took the time span of the nineteenth century, the growth of industry and the decline of agriculture to create new social and economic forces which, while removing one set of anachronisms, inevitably helped to destroy the other.

NOTES

1 See E. J. Hobsbawm, 'The Age of Revolution' (1964), p. 33: 'between 1760 and 1830 what emerged was... a class of agricultural entrepreneurs, the farmers and a large agrarian proletariat...'.
2 For a discussion on low wages and high rents see H. de B. Gibbins, 'Industry in England' (1896), p. 442.
3 Flora Thompson, 'Larkrise to Candleford' (1965), p. 191. 'Every morning... the Rector arrived to take the... children for scripture. To order myself lowly and reverently before my betters was the clause he underlined in the Church Catechism... God had placed them... in the social order... to envy others or change their own lot in life was a sin.'
4 'Church Times', 4 Feb. 1898, p. 117.
5 Gareth Stedman-Jones, Some Social Consequences of the Casual Labour Problem in London 1860-1890 With Particular Reference to the East End, PhD. thesis, Oxford, 1969, p. 14; On the Conditions of the Agricultural as Compared With the Manufacturing Population, JRASE, vol. VIII, p. 113: 'the employment of children in rural labour... deprives them... of the benefits of education. This... is a disadvantage by no means from their being agriculturists, but simply from their belonging to the lower classes... a disadvantage... from which they can never be freed in a country like this where the remuneration for a man's labour is never likely to rise greatly above that which is necessary to support himself, and those who are positively dependent upon him.'
6 C. E. Raven, 'Science, Religion and the Future' (Cambridge, 1968), p. 45.
7 See William Hoyle, 'Crime in England and Wales in the Nineteenth Century (1876), Table of Wage Rates in Industry in Manchester 1832, p. 79. Compare estimated average earnings for agricultural workers in Oxfordshire, 14s 6d, in Berkshire, 15s 11d, and in Buckinghamshire, 16s 4d, as late as 1902. (John Chapman, 'An Economic History of Modern Britain', vol. IV (Cambridge, 1951), p. 99.

8 Gilbert Phelps, 'A Survey of English Literature' (1965), pp. 235-6.

9 Karl Marx, 'Capital', vol. 1 (Moscow, 1958), p. 724, 'between 1801 and 1831... 3,511,770 acres of common land... were... [enclosed]'.

10 S. and B. Webb (eds.), 'The Public Organisation of the Labour Market, being Part Two of the Minority Report of the Poor Law Commission' (1909), p. 4: 'the famous "Allowance System" by which... [t]he farmers, secure of a constant supply of labour, lowered wages. The labourers, secure of subsistence, progressively lowered the quantity and quality of their effort.'

11 PP 1881 (c2778-II) XVI, RC Depressed Conditions, App. Q., XVIII. 'What has been the effect on the labourer of the adoption of labour saving machinery?' 'It had deterred him from striking' (Oxfordshire) (p. 340); 'Kept them in order' (Staffordshire) (p. 341); 'Would have been very great but for the union agitation and Mr. Arch. We are not quite so dependent upon them' (Warwickshire) (p. 342).

12 'Bucks Advertizer', 17 Aug. 1850, 'much of the wheat... beaten down... laid in many ways' (report from Berkshire); ibid., 12 Oct. 1850, 'on account of the bulk of the straw and all the corn being ready together every hand was eagerly engaged' (Mark Lane Express report on the Windsor and Eton area); 'Bucks Herald', 28 June 1856: 'Heavy crops of grass, hazardous weather and demand for labourers everywhere unite to make wages high... 6s per acre paid in Surrey for mowing natural grass... 6s to 8s per acre for clover'; 'Reading Mercury', 28 Aug. 1875: 'We would urge the advantage of stacking in the field rather than in the stack yard. Carting is saved... when labour is expensive and valuable less risk from accidental fire... stacks come sooner, and better into good condition in an exposed situation ('Ag. Gaz.' extract).

13 Ronald Blythe, 'Akenfield' (1969), p. 55. 'The holy time was the harvest... the farmer would call his men together and say "Tell me your harvest bargain" ' (John Grout, aged 88).

14 PP 1865 (3484) XXXVI, Inquiry on the State of Dwellings of Rural Labourers, Dr Hunter, App. no. 6, p. 251. 'It is the importunate policy of the settlement law to offer inducements to the demolishing cottages where they are urgently required, and to congregate them where they are not... farmers are so short sighted as to object to cottages on or near their farms, because they fear an augmentation of the poor rate (quoting Clare Sewell Read, On the Farming of Oxfordshire, JRASE, vol. XV (1854), p. 134); 'there are times when a farmer requires all the labour he can get; at others he thinks it proper to retain a very few hands only...'.

15 PP 1868-9 (4202-1) XIII, RC on Employment, App. pt. II Bj, p. 575 (24c).

16 Ibid., p. 574 (21a).

17 Ibid., p. 571 (8 and 8a).

18 'Labourers' Union Chronicle', 21 June 1873.
19 John Coker Egerton, 'Sussex Folk and Sussex Ways' (1892), p. 85.
20 Ruth W. How, 'Good Country Days' (1949), p. 135. 'It was their harvest... Often there was enough flour to last 'em through the winter.'
21 John Purse, unpublished MS, New Zealand, 1958, p. 27.
22 Ibid., p. 27, 'it was then that young men... went to the manufacturing towns thus relieving the village of its unemployed'.
23 PP 1893-4 XXXV, RC on Labour, App. B11, p. 218; 'North Wilts Herald' in October 1891 advertisements for farm servants; JOJ, 14 Oct. 1893: 'On Wednesday last Old Michaelmas Day, a number of agricultural employees with their goods and chattels passed through Burford en route for their field of labour for another year.'
24 Berks RO, C/ELS/1, Bradfield National School Log Book, 10 Oct. 1890. 'A number of families are changing this Michaelmas and we are losing a number of scholars in consequence...'; C/EL/10, Bradfield Village School Log Book, 11 Oct. 1899. 'We have this week lost sixteen children through Michaelmas removals [farmers changing hands]...'; ibid., 12 Oct. 1900. 'The usual removals have occurred this week causing a loss of six scholars.'
25 Blythe, 'Akenfield', p. 47. Fred Mitchell aged 85, Suffolk labourer who started work at 8 years old.
26 Cf. Fred Kitchen, 'Brother to the Ox' (1940), p. 101. 'So that was how we carried on before the Great War. Then the Agricultural Wages Act [Government Production Act, 1917]... altered the custom of hired servants being tied for a twelvemonth. But note: the tied-cottage system still prevented effective independence.
27 Cf. Dr Augustus Voelcker, The Influence of Chemical Discoveries on the Progress of English Agriculture, JRASE, 2nd edn, vol. XIV (1878).
28 Cf. 'The Journals of George Sturt', Mackerness, E. D., (ed.) (Cambridge, 1967), vol. 1, p. 38. 'In the early years of the century a great mass of country literature appeared... some solely concerned to exploit the charm and "romance" of village life and legend...'
29 Marianne Farningham (Mary Ann Hearne), 'Harvest Gleanings' (1903).
30 Emile Durkheim, 'The Division of Labour in Society' (1968 edn), Preface to 2nd edn, p. 17.
31 Note: Some improvement in winter work followed the instigation of government-free loans to encourage land drainage in 1840. (9 & 10 Vic. C.101, an act 'To Authorise the Advance of Public Money... to promote the improvement of land... by Works of Drainage', continued by 10 Vic. C.11, 11 & 12 Vic. C.119, 12 & 13 Vic. C.100, 13 & 14 Vic. C.21, 14 & 15 Vic. C.91 and in 1856, 19 & 20 Vic. C.9 'To Amend the Acts relating to the

Advance of Public Money to promote the improvement of land.') E.g. on Lord Shaftesbury's Dorset estate from 1853 to 1862 three large drainage schemes were embarked on: 'Happy prospects of my drainage efforts: many labourers will be required' (Oct. 1853); 'though mismanagement and direct fraud led to prolonged litigation and heavy financial loss'. See J. L. and Barbara Hammond, 'Lord Shaftesbury' (1923) (reprint 1939), pp. 164-9.

32 PP 1866 (3379) XXV, RC Poor Law Board, App. to the Sixteenth Annual Rep. of the Poor Law Board, South Midland Division III, pp. 248-315. Numbers: Feb. 1862 - indoor 7,851, outdoor 70,233, total 81,824; Sept. 1862 - indoor 7,851, outdoor 61, 410, total 69,261; Feb. 1863 - indoor 10,934, outdoor 69,391, total 80,325; Aug. 1863 (3rd week) - indoor 7,660, outdoor 61,625, total 69,285.

33 'Labourers' Union Chronicle', 15 May 1875. A labourer, the privacy of whose cottage was invaded, objected, but was told 'I am so much above you - I am a clergyman and you are but a labourer.'

34 See Appendix C, pp. 199-200.

35 Cf. Thomas Mackay, 'History of the English Poor Law', p. 193. 'The Labourers are getting more independent and are moving farther afield' (at Bucklebury, Bradfield Union, Berks, 1835); ibid., p. 198. The labourer 'bestirs himself and carries his labour to the best market, even if he has to move to the next parish or county'.

APPENDIX A

On Shocking Corn

In setting up the sheaves the numbers put together in the stooks or shocks varied from six to 12 according to the custom in various parts of the country as did the terminology - stooking, shocking, stacking, sticking up meaning the same operation. Stooks or shocks of six sheaves, sometimes eight were most usual in Berkshire, Buckinghamshire and Oxfordshire - shocking up, stooking, sometimes sticking up, described the operation but stacking in these counties was essentially a term reserved for the work involved in rick building. Six sheaves put opposite each other, from north to south, and one at each end was the Worcestershire practice. The general custom in Wiltshire was to set up the sheaves in double rows, usually ten sheaves together, traditionally for the convenience of the tithing man; in Yorkshire the sheaves were set up in stooks of 12. The number of sheaves in Shropshire, however, varied - eight, ten or 12 formed a 'mow', sticking up was the local term and if the weather was doubtful two of the sheaves were placed ears downwards at the top of the mow for covering or 'hooding' it. No hooding occurred in Hereford. The sheaves were merely 'stacked', while by 1867 the Warwickshire practice of 'capping' (hooding) was no longer so widespread. Hooding was in fact only normal practice in wet districts. In Cheshire the 'hattocks' of ten sheaves included two hooders carefully put on top; in Derbyshire 'stacks' of 12 sheaves were made - two in the middle, four on each side, two on the top with butt ends meeting and both sheaves tied together. The form of stook devised by J. Eames of Sussex, a farmer in the 1860s, who claimed it seldom blew down and resisted the wet better than any other, was put on the land (the highest point) and the eight sheaves in the shock were arranged thus: four sheaves were first put up so as to form a square, then another four were added all brought together at the top as much in a cone as possible. When well set up each side of the shock presented a straight line of three sheaves. Loss of grain could come from rough or careless handling when shocking especially when the corn was ripe or over ripe. If the head of the sheaf was grasped, or handling was too near the ears of corn, there was considerable risk of shedding or of broken ears; band snatching, another purchase point, risked breaking open the sheaf. Ideally a sheaf was picked up at a point along the sheaf so that when lifted, the butt a little outweighed the head of grain. Shocking was usually undertaken by a pair of workers who with a sheaf thus picked up in each hand and with quick swinging movements, brought all four sheaves upright and propped them against each other with a firm bumping of the butts. For the expert there was no call for the admonition 'Bump 'e down hard on t'arse, man!'[2]

Involving continuous stooping and lifting, shocking was never more arduous than in the middle and later years of the nineteenth century, for greater fertility had led to heavier grain yields and greater straw length, the small sheaf associated with high reaping was fast disappearing, and short-strawed varieties of wheat and oats were still in the future. Apart from extra weight lengthy straws produced cumbersome sheaves. This was an aspect of the wheat crops grown on Lord Wantage's Berkshire estate about Lockinge until the early 1890s. White wheats were planted, partly for the quality and quantity of their straw, which sold at £10 per ton to the Luton straw-hat makers. Only when the market dried up, were higher grain-yielding red wheats grown.[3]

NOTES

1 Mainly based on Harvest Work and Wages, 'Ag. Gaz.', 24 Aug. 1867, pp. 888-91, and personal experience.
2 G. E. Moreau, 'The Departed Village' (1968), p. 77.
3 See F. H. Burch, Lady Wantage's Berkshire Estate, 'The Estate Magazine', vol. VIII (1908), pp. 433 ff.

Appendix B

On 'Harvest Beer' and the Temperance Movement

Temperance reformers often organised harvesting competitions between teetotal and drinking harvesters, in the hope that this would convince farmers that drink did not get in the harvest more quickly. Mr Webb, a Reading farmer and 33 years a teetotaller, told a temperance meeting in May 1837 that he had tried to induce his employees to work on teetotal principles for a higher wage, but they had abandoned the attempt because of the ridicule they faced from other harvesters.[2] A more successful attempt was made by Henry J. Wilson who farmed 1,000 acres at Newlands, near Mansfield, Notts, in 1854.

> It is usual to furnish malt liquor at hay and corn harvest, to meet the need for extra exertion at these seasons. This frequently consists of a definite quantity [about five pints] of ale, and an indefinite quantity of small-beer - the daily value being from 1s to 1s4d per day. If 5s a day be a fair average of the daily wages, then we find the hard-working harvestman receiving one-fifth of his pay in drink; drink, too, which actually prevents his doing as much work as he otherwise would do. The question will here naturally arise, 'What is the best substitute?' 'Water, pure water', will be the answer of those who never did a hard day's work in harvest, but they will not be supported by many who have had that experience... In my own experience, tea and coffee have been found most acceptable and... we have gone on steadily for eight years [1862] with no difficulty in procuring plenty of help, even in harvests when labour has been scarce. I have never had a man, not even a stranger, turn his back on an offer of work on account of the drink. The same gangs from a distance will return year after year in the hope of getting employment.[3]

'Harvest beer', however, was one of several factors which prevented the nineteenth-century temperance movement from ever being really secure in country areas. Throughout the nineteenth century they tried to attack the notion that alcohol strengthens the physique, and that fatness is an indication of strength. Elibu Burritt reports that Samuel Jonas, the great farmer near Royston, was in 1863 paying his men in cash instead of beer.[4] Similarly, J. Arnot of Carshalton, Surrey, maintained that he got more work done and that the men at any hour of the day were fit to be spoken to and always civil, but normally a cash payment instead of the provision of beer only meant labourers bought their own;[5] see pp. 101-19 for the place of beer in the complicated structure of harvest payments. Many pamphlets were written by teetotallers against harvest beer[6] and the pressure of the Church of England Temperance Society helped to ensure that payment in beer was forbidden under Acland's Agricultural Wages Bill of 1887.[7]

NOTES

1 Sources and comments mainly Dr B. Harrison's, Corpus Christi College, Oxford, with some additions.
2 'London Temperance Intelligencer', 3 June 1837, p. 239.
3 Dietary of the British Labourer, JRASE, vol. XXIV (July 1863).
4 Elibu Burritt, 'Walk from London to John O'Groats' (1864), p. 175.
5 'Ag. Gaz.', 24 Aug. 1869, p. 891, col. 1.
6 E.g. The anon pamphlet Intemperance: Its Bearing Upon Agriculture, Pusey House, n.d., pamphlet 2235. The report of the York Convocation Committee on Intemperance (1872), p. 4 also condemned the practice. See also Thos. Snow's pamphlet 'Hard Work in the Fields in Hot Weather' (1877).
7 On this latter measure, see P. T. Winskill, 'The Temperance Movement & its Workers', vol. IV (1892), p. 150 and G. W. Hilton, 'The Truck System' (1960), p. 143; the practice in fact did not cease (see p. 112). As for the incidence of the practice, see A. Tindal Hart, 'The Country Priest in English History' (1959), p. 110, the anti-Corn Law League periodical 'The League', 2 Dec. 1843, p. 157, and anon, 'Essay on the Repeal of the Malt-Tax' (Total Repeal Malt-Tax Association, 1846), pp. 11-14.

Appendix C

Settlement, Removal and Mobility

The right to parochial relief until 1834 was dependent on a settlement acquired by either parentage, birth and marriage or alternatively by hiring and service, apprenticeship, or purchasing or renting a tenement. Unsettled labourers were liable to be removed to the place where last they had claim to a settlement but only if they had become chargeable (1795 35 Geo. III C.101). If through sickness or accident a labourer was removed but on return subsequently sought relief, only a certificate of settlement procured on his first removal made certain that he would not be treated as a vagrant incurring a maximum penalty of one month imprisonment or a fine of £5 (1824 5 Geo. IV C.83 S.3 Vagrancy Act). To prevent settlement being acquired through hiring a current practice was to engage labour for either less or more than the qualifying period (one year); for example, 'no settlements were gained by male servants in closed villages... [T]here are no records of settlements obtained by servants in Tysoe.'[1] An evasive practice existed in Oxfordshire - to date consecutive hirings from Old Michaelmas (11 October) to New Michaelmas (29 September), then from New Michaelmas to Old Michaelmas seriatim.[2] The elimination of settlement acquired by hiring under the new Poor Law, therefore, less affected settled labourers seeking other employment than ended an opportunity for unsettled labourers to acquire a new settlement. In the circumstances of 1834 the threat of removal was seen as a pertinent influence on unsettled labourers. 'At present they are confessedly superior, both in morals and industry to those... settled in the parishes in which they reside,' was the opinion which also influenced the Poor Law commissioners to reject the idea of settlement by residence: 'Make... residence a settlement and they will fall back into the general mass.'[3] Not until the Divided Parishes Act, 1876 (39 & 40 Vic. C.61) enacted that continuous residence of three years would confer a settlement did the Settlement Law directly influence the decline in removals. This occurred in the interim period indirectly since the granting of non-resident relief continued to be widely adopted: '[A]t Lady Day 1846 there were 12,249 persons receiving [it]... prevent[ing] some 32,899 warrants of removal... a less costly expedient...'[4] From this date a new status of 'irremovability' imposed a bar against its cruelty dependent on five years residence (9 & 10 Vic. C.66), and reduced it to three (1861 24 & 25 Vic. C.55) and subsequently to one (1866 28 & 29 Vic. C.79 Union Chargeability Act).[5]

NOTES

1 See A. K. Ashby, 'Poor Law Administration in a Warwickshire Village' (1912), p. 67.
2 Cf. H. R. Wright and T. E. Lones, 'British Calendar Customs', vol. III (6 vols., The Folklore Society, 1940), p. 95; see also E. J. Hobsbawm and George Rude, 'Captain Swing' (1968), p. 46, 'got round by some small legal trick... employing a man for only fifty weeks in the year'.
3 See Thomas Mackay, 'History of the English Poor Law' (1896), pp. 349-50.
4 Ibid., p. 350.
5 For incidence of removal see S. and B. Webb, 'English Poor Law History' (vol. VII, English Local Government, 1963 edn.), p. 410-34. Footnote p. 434: 1841 - 8,412 orders of removal; 1849 - 13,867 orders relating to 40,000 persons; 1851 - 30,000 cases of actual removal; 1882 - 4,211 orders relating to 6,233 persons with 2,692 more removed without formal orders; 1895 - about 6,000 persons removed; 1907 - upwards of 12,000 persons 'more from London... than from Rural Unions'.

Appendix D

Prices of Beer 1858 and 1889 and Some Consumer Prices

Beer 1858	£	s	d	
Brown Stout	1	6	0	the Kilderkin (Cask 16 gallons)
Porter	1	2	0	ditto
Strong Beer		1	6	per gallon
Half and Half		1	3	ditto
Extra Ale		1	0	ditto
Ale			10½	ditto
Second ditto			8	ditto
Third ditto			6	ditto
Table ditto			4	ditto
Beer 1889				
Light Dinner Ale		7	6	per gallon (9)
Bitter		9	0	ditto
IPA		12	6	ditto

Sundry Clothing[1]	
Men's suits	50s to 80s
Trousers	13s 6d to 22s 6d
Ready-made Suits	18s 6d
Trousers	10s 6d
Summer dresses	7½d-1s 6d yd

Note:[1] Bedsteads, 10s 11d-17s 11d; Mattresses, 13s 11d.
Sources: Price List from W. Miller Eagle Brewery, New Road, Oxford (Bod. Alm. Oxon, 8s 1038); OT, 3 Aug. 1889, p. 2 (advertisement page).

GLOSSARY

BERKSHIRE

arnest/ earnest (Oxon)	The 'earnest' or 'arnest' money is a shilling given on hiring a servant; it completed the contract.
aftermath/ lattermath	The second crop of grass, i.e. aftermowth.
barleyoleys	The beards of barley.
bin	The corn chest in the stable (always secured by a padlock).
boddyhoss/ body horse	The horse of a team next in front of the 'thiller' (see below).
curvew bell	Still rung at Blewbury between Michaelmas and Lady Day (1888).
dummle	In corn or hay, damp.
ell raayke/ hel-rake (Oxon)	The large sized rake used for raking hay left behind where 'cocks' have been 'pitched' into the waggon.
graains	The forks of a prong; thus a dung prong is a three graained prong. (Malt after all the goodness is extracted in brewing - brewer's grains.)
grist	Corn brought to the mill for grinding.
haam/haulm	Stubble or straw of vetches peas and beans. The 'haam' rick in the Vale of Berkshire is of beans or wheat straw - vetch 'haam' rick in the hill part of the county.
hack	To fag or reap vetches peas or beans.
harvest wheam	The festival which winds up harvest work.
kern	Corn.
ley	Growing grass.
lot	The feast time at some villages - Drayton 'lot'.
mayster/me-uster	Farmer by his men.
me-ad	A meadow.
mow/mou (Oxon)	Corn or straw stacked in a barn.
mullock	Wet straw.
pollards	The ground husk of wheat, medium size; the coarsest size being 'bran' and the finest 'topping'.

prong	The metal part of the implement for moving hay, straw etc. The wooden part is the 'prong-handle'. The ordinary prong has two forks, whilst the dung prong has three.
rack-up	To close the stables for the night after littering the horses and giving them their 'vead'. 'Rackin'-up' time marks the conclusion of the day's work for carters and carters' boys.
revel	An annual village merry-making as Chapel-Row 'Revel'.
rick/wrick	'Rick' is always used for stack, 'hay rick', barley rick etc. A rick cloth is a waterproof sheet placed over the top of a rick to keep out the wet until such time as the rick may be thatched.
rick yard	Place where ricks are made.
rip-hook	A sickle.
scoop	A wooden shovel as used for shovelling corn after it is threshed.
shock	A few sheaves of corn placed together in the field so that the ears and straw may dry in the sun before the rick is formed.
shock-up	To form sheaves into shocks.
strapper	A journeyman labourer coming for work at harvest time or hay making.
stubs	Stubble. A field lying in stubble is called a 'po-us o' whate-stubs' or a 'pe-us o' wut-stubs' (oat).
taayke-in	To 'taayke-in' a rick is to thresh out the corn.
thiller/viller	The shaft horse of a team.
transmogrivied	Transformed in appearance, disguised.
tuck	To trim. A rick is said to be 'tucked' when raked down so as to take off loose surface straws and leave the others nearly lying in the same direction.
upperds/ uppards (Bucks)	Upwards.
vaairin	A present bought from a country fair by one who is fortunate enough to go, to another obliged to stay at home.
vhayle	The country along the Thames Valley as about Blewbury, Hagbourne, Moreton, Didcot etc. is so called. The other part of the country is styled the 'Hill Country'.
ve-ast	The annual village merry-making, usually held on the Dedication Day of the parish church, thus Hagbourn Ye-ast (see also lot and revel).
vogger	A farmer's groom - responsible for feeding pigs and cattle.
vorruss	The leading horse in a team.
vraail	A flail.

wuts	Oats.
yarnes	'Yarnest' money.
yead-lan	A headland. Part ploughed at the head or top of the main plough.
yecker	An acre.
yelm	To straighten straw in readiness for threshing.
zwathes	Rows of hay when first cut.
zwingel	The top part of the threshing flail.

OXFORDSHIRE

backaive	To winnow corn through a fine sieve, called a backheaving sieve.
belten	A large bundle of straw.
blind	When spring corn does not thrive or grow well, it is said to 'look very blind'.
crass crappin	Sowing the crops out of the accustomed order.
earnest money	A shilling given at a hiring fair to a servant to 'bind the bargain'.
faggin	Cutting corn with a heavy hook and a hooked stick called a fagging stick.
four o'clock	A tea in the hay field - Holton, Islip, Yarnton.
frail	A flail.
galley hill	The usual name for Witney Union which is situated on Galley Hill.
grinsard grounds	Pasture fields.
haams	Stubble cut after the corn is carried - straw of white crops only.
hackle	To rake hay into rows after it has been 'tedded': usually called to hackle in or up. (To leet - Holton.)
hadlans	The top and bottom lands which are ploughed the reverse way to the others.
hayn up	To reverse grass for hay.
land	A ridge and furrow.
muster	Mister.
maerster	Master (Islip).
old standards	Natives of a place. 'I and Master Viner be the uny two old standards left.'
rakers-arters	The people who rake after the waggons in the hay field.
run-away-Mop	The third Mop or hiring fair said to be composed of servants who have been hired at the previous fairs and have run away from their situation.
tie up	To bind up sheaves of corn in a band (called a bond) of corn stalks. 'My ole dooman's a gwain tiein up fer ma [me].'

NOTE

1 Sources: B. Lowlsey, 'Berkshire Words' (1888) and Mrs Parker, 'Supplement to Glossary of Words in Oxfordshire' (1881)

BIBLIOGRAPHY

MANUSCRIPTS

Bodleian Library, Oxford (Bod.)

Notes of Thomas Banting of Filkins, TOP. OXON C220
Miscellaneous papers relating to Bicester, TOP. OXON. d.199
L. N. Letch, Bucknell Village (1933), Number of additions 1935-6. MS TOP. OXON. C.515
Manning (Percy) paper, G.A. OXON 4°186
C. E. Prior - miscellaneous papers relating to Oxfordshire, R. TOP. 330.

Clare College Library, Cambridge

Cecil Sharpe Collection

Berkshire County Record Office, Reading

Log Book, Bradfield National School, 1866-93, C/EL9/1
Log Book, Bradfield Village School, 1886, C/EL10
Minute Book, East Ilsley School Board, C/EB5

Buckinghamshire County Record Office, Aylesbury

Log Book, Addington Village School, 1872-1915, E/LB1/1

Oxfordshire County Record Office, Oxford

Dillon Papers: Farm accounts and Labour Books 1757 (DIL)

East Sussex County Record Office, Lewes

Log Book, 1863-95, Stanmer and Falmer National School, E 237/12
Log Book, 1875-(1900), East Dean and Friston School, E.304/25/1
Rough Minute Book, 27 Apr. 1876 - 31 May 1883, Boro' of Rye School Board, E.467/1/4
Battle Abbey Estate - Account of Labour and Team Work, 6 May 1839 - 16 May 1840 (Ashb. Ms. 2009); 8 Apr. 1882 - 26 Jan. 1883 (Ashb. Ms. 2012)

University of Reading Library Accessions of Historical Farm Records

This collection of farm accounts, labour books, farm diaries and sundry papers has been extensively consulted; those mainly referred to in the text are listed below.

Berkshire

H. W. Cozens, Wittenham Clump Farm, Little Wittenham, BER 10/1
J. A. Badcock, Wick Farm, Radley, Account Book, BER 13/3/6, Labour Books, BER 13/5/1-6
Steventon Labour Book, May 1846-Jan.1848, BER 16/2/1
Park Farm, Wallingford, Cropping Register, 1894-1932, BER 19/7/1

Bradley Farm Estate Chievely, Books of Accounts, 1 May 1861 - 28 May 1868, 2 May 1868 - 17 Dec. 1875, 18 Dec. 1875 - 6 Aug. 1884, 19 Mar. 1891 - 8 Dec. 1899, BER 28/1/1-4
Labour Book, 23 Aug. 1867 - 15 Oct. 1869, BER 28/3/1
Manor and Grazeley Farms, Account Book, 15 Aug. 1895 - 11 Aug. 1902, BER 29/1/1

Buckinghamshire

Refill, Manor Farm, Horton Labour Books, BUC 1/5/1-3
Holmes Hill Farm, Stokenchurch Farm, Account Book, 25 Aug. 1849 - 12 Oct. 1866

Oxfordshire

John Barford Fawcett, OXF 2/2/4
Manor and Church Farms, Aston Bamford, OXF 6/1/3
Holton, Sibford Ferris - Union wage claim, OXF 14/3/5
Nettlebed Estate, OXF 18/3/1

Cambridgeshire

Hall Farm, Great Wilbraham, Underwood Hall and Partridge Hall Farm - two harvesting agreements for 1891 and 1893, CAM 1/1/1-2

Dorset

Monkton Farm, Tarrant Monkton, BER 20/2/1

Gloucestershire

Joseph Martin, Ripple Tewkesbury, GLOS 2/2/2

Huntingdonshire

W. B. Fowler, Lodge and Common Farms, North Farm, Leycourt Farm, Hardwick Farm, Manor and Woodhams Farms - Labour Books, 9 Aug. 1897 - 21 Aug. 1916, containing inscribed harvest agreements for 1897, 1898, 1903 and 1904, HUN 2/1/1

Wiltshire

Thomas While, Castle Hill, Cricklade - Contract of Hiring, 11 Oct. 1911, OXF 6/3/1
Stratton, Kingston Deverill Farm, Farm Diaries 1867-9, WIL 6/4/1-3
Farming History of the Stratton Family 1864-1933 (William Stratton), WIL 6/8/1
Log Book 1868 - Wall St Village School, nr Lichfield, Staffs (uncatalogued Reading University Library Archives)

Privately Printed Reminiscence

John Purser (b. 1878) of Ilmington, nr Shipston on Stour, Warwicks, (Wellington NZ c. 1958). Began farm work at the age of 13 'leading' horses: emigrated to New Zealand in 1899 after a short time working on the railways; three older brothers had all earlier found alternative employment - one joining up, the other two leaving home to work in 'a distant town'.

Recorded Interviews (Personal Communication)

Mrs Kerry Beaumont, Cottages, Headington Quarry, Oxford to B. Reaney, Corpus Christi College, Oxford (9 Aug. 1969). Headington Quarry villagers to Raphael Samuel, Ruskin College, Oxford (1969-70). Walter R. King (b. 1895) Library Farm, Garsington to David H. Morgan (11 Nov. 1969)

CONTEMPORARY NEWSPAPERS AND PERIODICALS

Local

Berkshire
'Berkshire Chronicle'
'Reading Express'
'Reading Mercury'
'Reading Observer'
'Reading Standard'
'Wantage Free Press'
'Windsor and Eton Express'
Buckinghamshire
'Aylesbury News'
'Buckingham Express'
'Bucks Advertizer'
'Bucks Herald'
Oxfordshire
'Banbury Chronicle'
'Banbury Guardian'
'Bicester Herald'
'Jackson's Oxford Journal'
'Oxford Chronicle'
'Oxford Times'
Warwickshire
'Leamington Spa Courier'
'Royal Leamington Spa Chronicle'
'Warwick Advertizer'

National

'Agricultural Gazette' (1844), subsequently 'Gardeners' Chronicle and Agricultural Gazette' (especially 1850, 1860, 1867 and 1868)
'All The Year Round'
Pinchback's Cottages
Pinchback's Amusements
The Farm Labourer's Income
Three short unsigned sketches, vol. VII (1862)
'The Church Times'
'Cornhill Magazine'
The Life of a Farm Labourer, vol. X (1864)
Harvest, vol. XII (1865)
The Irish, vol. XII (1865)
The Agricultural Labourer I, vol. XXVII (1873)
The Agricultural Labourer II, vol. XXVII (1873)
'The Countryman'
Peasemore, Berks, vol. 56 (1959)
A Strike of Farm Workers, 18 Apr. 1866
'The Estate Magazine'
F. H. Purchas, Lady Wantage's Berkshire Estate, vol. VIII (1908)
'Illustrated London News'
'Implements Manufacturers' Review'
'The Justice of the Peace'
Questions and Answers on Settlement, vol. XXVII, no. 1 (1863); vol. XXXVIII, no. 11 and no. 29 (1874); vol. LXII, no. 2 (1898)
'Labourers' Union Chronicle' (1872)
'The Land Worker'
'Morning Chronicle' (especially Sept. 1849 - Mar. 1850)

'The Nineteenth Century'
Hodge and His Parson, vol. XXXI (1892)
Hodge at Home, vol. XXXI (1892)
'Punch'
'Quarterly Review'
Agricultural Gangs, vol. 123 (1867)
The Lock-out of the Agricultural Labourers, vol. 137 (1874)
'Westminster Review'
Agricultural Depression Unmasked, vol. CXLIV (1895)
The New Minister of Education and His Work, vol. CXLIV (1895)

PARLIAMENTARY PAPERS

Select Committee (House of Lords) Report on Agriculture 1837 (464) V
Select Committee Report on Machinery 1841 (201) VII
Select Committee Report on Wastelands, Labouring Poor (Allotment of Land) 1843 (402) VII
Poor Law Commissioners' Reports on Employment of Women and Children in Agriculture 1843 (510) XII
Select Committee Report on Andover Poor Law Union 1846 (663-1) V
Reports to the Poor Law Board. Settlement and Removal of the Poor 1850 (1152) XXVII, 1851 (675) XXVI
Supplement to Mr. Coode's Report on Law of Settlement and Removal 1854 (493) LV
Agricultural Statistics 1854. Poor Law Inspectors Reports Wastelands 1854-5 (1928) LIII
State of Popular Education in England 1861 (2794-I) XXI
Royal Commission on Public Health (3416) XXVIII
Dwellings of the Rural Labourers. Dr H. J. Hunter's Report (7th Report Board of Health 1864) 1865 (3484) XXVI
Poor Law Board Report 1866 (3379) XXV
Workhouses. Reports by Inspectors in their Respective Districts 1867-8 (35) LXI
Royal Commission on Employment of Children, Young Persons and Women in Agriculture 1867, 1867-8 (4068) XVII, 1868-9 (4202) XIII, 1870 (c70) XIII
Report on Inclosure Act 8 and 9 Vic. C. 118 (whether provisions for the labouring poor have been carried out) 1868-9 (304) X
Royal Commission on Master and Servant Act 30 & 31 Vic. C.141 and the Criminal Law Amendment Act 34 and 35 Vic. C.32 (Labour Law Commission), 1874 (C.1094) XXIV, 1875 (C.1157) XXI
Royal Commission on Depressed Conditions of the Agricultural Interests, 1881 (C.2778) XV (C.2778-II) XVI (C.3096) XVII, 1882 (3309) XIV (3375-I) XV
Royal Commission on Education 1888 (c.5469) XXXVIII
Abstract of Royal Commission on Agricultural Interests 1879-1882, 1893-4, XXVII
Royal Commission on Labour: The Agricultural Labourer 1893 (C.6894-I) XXXV
Royal Commission on Agricultural Depression 1894 (C.7400) XVI, 1895 (C.7915) III (C.7691) XVI, 1896 (C.7981) XVI (C.8021) XVII, 1897 (C.8540) XV
Reports by Mr Wilson Fox on the Wages Earnings and Conditions of Employment of Agricultural Labourers in the United Kingdom, 1900 (CD346) LXXXII, 1905 (CD 2376) XCVII (Appendix V - Migratory Labourers in England and Wales)
Report on the Decline in the Agricultural Population of Great Britain 1881-1906 (CD 3273) XCVI
'The Public Organisation of the Labour Market, being Part II of the Minority Report of the Poor Law Commission', S. and B. Webb (eds.) (1909), p. 4.
Agricultural Statistics, 1900. Migratory Agricultural Labourers.

Accounts and Papers

Judicial Statistics, C854 (2692) LXIV
Law: Police, 1856 (441) L
Statistical Tables 1857-8 LVII
Report: Poor, United Kingdom - Able Bodied Paupers (240), Agricultural Parishes (151) 1865 XLVIII
Select Committee Report: Contract of Service between Master and Servant, 1865 (370) VIII, 1866 (449) XIII
Miscellaneous Statistics, 1866 LXXIV
Law and Crime, 1873 (385) LIV, 1874 (360) LIV
Agricultural Statistics: Ireland 1884 (C-4150) LXXXV
Statistical Tables and Report on Trade Unions. 4th Report 1889-1890, 1890-91 (C6475) XCII
Report of Labour Department; Strikes and Lockouts 1893-4 (C7900) XCII
Departmental Committee Report: School Attendance and Child Labour, 1893-4 LXVIII
Report on Migratory Agricultural Labourers 1900 (CD 341) C1
Agricultural Statistics Ireland 1902 - Irish Migratory and Other Labour 1902 CXVI
For agricultural returns see A & P under either Agricultural Statistics or Miscellaneous Statistics.
For offences against the Master and Servant Acts (1824 4 Geo. IV C.34, 1867 30 & 31 Vic. C.141) and The Employers and Workmen Act (1875 38 & 39 Vic. C.90) see Judicial Returns (1857 -) or Miscellaneous Statistics - breach of contract listed as one of eight separate offences under Labour Laws 1894 -; similarly for offences against the Education Acts 1870 33 & 34 Vic. C.75, 1876 39 & 40 Vic. C. 79, 1880 43 & 44 Vic. C.23, 1893 56 & 57 Vic. C.51 and 1899 62 & 63 Vic. C.32.

CONTEMPORARY PRINTED SOURCES (INCLUDING ARTICLES UNDER AUTHOR)

Ackland, Thomas Dyke, On Lodging and Boarding Labourers as practised on the farm of Mr. Sothern, M.P., JRASE, vol. X (1849)
Andrews, W. 'England in Days of Old' (1892)
Anon, On the Condition of the Agricultural as Compared with the Manufacturing Population, JRASE, vol. VIII (1847)
—— 'The Harvest Festival at Lilbrook' (Pamphlet, 1861), Bodleian Library 1419 f 1316 (33)
—— 'Stubble Farm' (1880)
—— 'The Farm Labourer's Catechism' (Pamphlet, Andover, 1884), Bodleian Library 23214 C.11(4)
Arch, Joseph, 'The Story of his Life' (1898)
Atkinson, J. C. 'Forty Years in a Moorland Parish' (1891)
Baring-Gould, S. 'The Vicar of Morwenstow' (1876)
—— 'An Old English Home' (1898)
—— 'Old Country Life' (1899)
The Labourer's Friend. A Selection from the Publications of The Labourer's Friend Society, (1835), Efforts to provide labourers with allotments.
Beasley, A. 'A History of Banbury' (1841)
'Berkshire Notes and Queries' (1890)
Besant, W. 'The Eulogy of Richard Jefferies' (1888)
Best, Henry, 'Rural Economy in Yorkshire in 1641' (Surtees Society Publication, vol. XXXIII, 1857)
Blakeborough, Richard, 'Wit, Character, Folklore and Customs of the North Riding of Yorkshire' (1898)
Bonser, W. Bibliography of Folklore, 'Folk-lore', vol. CXXI (1961), Index of Folk-lore, vols. 1-68 (1890-1957)
Booth, Charles, 'The Aged Poor in England and Wales' (1894), especially pt. iv Old Age in Villages
Bourne, George, (Sturt) 'The Bettesworth Book' (1901)

—— 'Memoirs of a Surrey Labourer' (1907)
—— 'Change in the Village' (1912)
—— 'Lucy Bettesworth' (1913)
—— 'William Smith, Potter and Farmer 1790-1858' (1920)
—— 'A Farmer's Life' (1922)
—— 'A Small Boy in the Sixties' (Cambridge, 1927)
Bowley, A. L. The Statistics of Wages in the United Kingdom during the Past Hundred Years, JRSS, vol. LXII, pt. III (1899), p. 555
Brand, John, 'Observation on the Popular Antiquities of Great Britain', vol. II, 132nd edn (1849)
Britten, J. 'Old Country and Farming Words Gleaned from Agricultural Books' (English Dialect Society, 1880)
Burgess, J. Tom, 'Life and Experiences of a Warwickshire Labourer' (1872)
Caird, J. 'High Farming Vindicated and Further Illustrated', 2nd edn (1850), Oxford Union Library Pamphlet Collection, vol. 26.
—— 'English Agriculture in 1850-51' (1852)
—— General View of British Agriculture, JRASE, 2nd Ser., vol. XIV, pt. II (1878)
—— 'The Landed Interest' (1898)
Chambers, FRS, E. 'Encyclopaedia', vol. III (1779)
Chapman, D. H. 'A Farm Dictionary' (1853)
Clarke, John Algernon, Practical Agriculture, JRASE, 2nd Ser., vol. XIV, pt. II (1878)
Claydon, Arthur, 'The Revolt of the Field' (1874)
Clifford, Frederick, 'The Agricultural Lockout 1874' (Edinburgh, 1875)
Clutterbuck, J. C. On the Agriculture of Berkshire, JRASE, vol. XXII (1861)
—— On the Farming of Middlesex, Prize Essay, JRASE, 2nd Ser., vol. V (1869)
Cobbett, William, 'Rural Rides' (1830) (Intro. Geo. Woodcock 1967 edn)
Cocks, A. H. 'The Church Bells of Buckinghamshire' (1897)
Connell, Sir Isaac, 'The Farmer's Legal Handbook' (Edinburgh, 1894)
Cooke, G. A. 'Oxfordshire' (c. 1822)
Davidson, John, 'A Random Itinerary' (1893)
Davies, David, 'The Case of Labourers in Husbandry' (1795)
Davies, Maud F. 'Life in an English Village' (1909), Parish of Corsley, Wilts
Davis, R. (of Lewknor), 'General View of the Agriculture of Oxfordshire' (1794)
de Lavergne, L. 'The Rural Economy of England, Scotland and Ireland' (1855)
Dent, J. D. Agricultural Notes on the Census of 1861, JRASE, vol. XXV (1864)
—— The Present Condition of the Agricultural Labourer, JRASE, 2nd Ser., vol. VII (1871)
Denvir, John, 'The Irish in Britain' (1892)
Ditchfield, Peter Hampson, 'Bygone Berkshire' (1896)
—— 'Old English Customs Extant at the Present Time' (1901)
—— (ed.) 'Memorials of Old Oxfordshire' (1901)
—— (ed.) 'Memorials of Old Buckinghamshire' (1903)
—— 'Cottage and Village Life' (1912)
Dudgeon, John, On British Agricultural Statistics and Resources in Reference to the Corn Question, 'Quart. Journ. Agric.', vol. XII (1842)
Eddison, Edwin, Harvesting in a Bad Season, JRASE, vol. XXIII (1862)
Edwards, George, 'From Crow Scaring to Westminster' (1922)
Egerton, John Coker, 'Sussex Folk and Sussex Ways' (1892)
'Encyclopaedia of the Laws of England', vol. VIII (1898)
Engels, Friedrich, 'The Condition of the Working Class in England' (1846), Trans. W. O. Henderson and W. H. Chaloner (1958 edn)
'English Dialect Dictionary, The' (1900)
Falkner, J. M. 'A History of Oxfordshire' (1899)
Farningham, Marianne (Mary Ann Hearne), 'Harvest Gleanings' (1903)
Fletcher, Joseph Smith, 'From the Broad Acres' (1899)
'Folk-lore', vol. I (1890)
'Folk-lore Journal', vol. I-VII (1883-9)
'Folk-lore Record', vol. I-V (1878-82)

Fowler, John Kersley, 'Echoes of Old Country Life' (1892)
—— 'Records of Old Times' (1898)
Fox, Arthur Wilson, Agricultural Wages in England and Wales during the Last Fifty Years, JRSS, vol. LXVI, pt. II (1903)
Garnier, Russell M. 'Annals of the British Peasantry' (1908)
Gatty, Rev. Alfred, 'The Bell, its Origin, History and Use' (1848)
Gibbins, H. de B. 'Industry in England', (1896)
Gibbons, Agnes, and E. C. Davey 'Wantage Past and Present' (1901)
Gibbs, J. Arthur, 'A Cotswold Village' (1898)
Gibbs, Robert, 'A Record of Local Occurrences' (4 vols., 1840-1880, Aylesbury, 1882)
—— (ed.), 'Bucks Miscellany' (1891)
Gomme, Alice B. Notes on Berwickshire Harvest Customs, 'Folk-lore', vol. XIII (1902)
Green, J. L. 'Rural Industries of England' (1892)
Greenwood, James, 'On the Tramp' (1883)
Griffith, G. 'Midland Counties Reminiscences' (1880)
Haggard, H. Rider, 'A Farmer's Year' (1899)
—— 'Rural England' (2 vols., 1903) (1906 edn), agricultural and rural researches carried out in 1901 and 1902
Hannan, John, On Reaping Wheat, No. 1, 'Quart. Journ. Agric.', vol. XII (1842)
—— On Reaping Wheat, No. 2, 'Quart Journ. Agric.', vol. XIII (1843)
—— Report on the Implements at the Royal Agricultural Society's Show at Liverpool; and on the Trials of the Self-binding Reapers at Highburth, JRASE, 2nd Ser., vol. XIV, pt. i (1878)
Hasbach, W. 'A History of the English Agricultural Labourer', Studies in Economic and Political Science, vol. 15 (LSE, 1896) (1908 edn)
Heath, F. G. 'The English Peasantry' (1874)
—— 'Peasant Life in the West of England' (1880)
—— 'British Rural Life and Labour (1911)
Heath, Richard, 'The English Via Dolorosa, or Glimpses of the History of the Agricultural Labourer' (1884), Bodleian Library 23214 C 11 (2)
—— 'The English Peasant' (1893)
Henderson, Richard, 'Manual of Agriculture' (Edinburgh, 1877)
Henderson, R. Sturge, 'Three Centuries of North Oxfordshire' (1902)
Hissey, James J. 'Through Ten English Counties' (1894)
'History, Gazetteer and Directory of the County of Oxford' (Peterborough, 1852)
'History of the Year Oct. 1881 - Sept. 1882' (1882)
Hone, William, 'The Every-Day Book', vol. I (c.1826), vol. II (c.1827) vol. III (c.1827-8)
Hoskyns, Charles Wren, 'TOLPA or The Chronicles of a Clay Farmer' (1852)
Hoyle, William, 'Crime in England and Wales in the Nineteenth Century' (1876)
Hulme, Frederick Edward, 'Proverb Lore' (1902)
James, William and Jacob Malcolm, 'General View of the Agriculture of Buckinghamshire' (1794)
Jefferies, Richard, 'Chronicles of the Hedges and Other Essays' (1879), Samuel J. Looker (ed.) (1948 edn)
—— 'Green Fernel Farm' (1880)
—— 'Hodge and His Masters' (1880) (rep. in 2 vols., 1966 with intro. by R. Williams)
—— 'Round about a Great Estate' (1880)
—— 'Nature near London' (1883)
—— 'Amaryllis at the Fair' (1887)
—— 'The Toilers of the Field' (1892)
—— 'Wild Life in a Southern County', illust. G. E. Collins (1937)
—— 'Field and Hedgerow' (1948)
—— 'The Nature Diaries and Note Books', Samuel J. Looker (ed.) (1948)
Jekyll, Gertrude, 'Old West Surrey' (1904)
—— 'Old English Household Life' (1925)
Jerrold, Walter, 'Highways and Byways in Middlesex' (1904)
Jesse, Edward, 'New Historical and Descriptive Survey of the County of Berkshire', illust. (1840)

——— 'Favourite Haunts and Rural Studies' (1847), Windsor and Eton district
——— 'Scenes and Occupations' (1853)
Johnston, William, 'England as it is... in the Middle of the 19th Century' (2 vols. 1851)
Keary, H. W. Report on the Farm Prize Competition (Oxfordshire Farms), JRASE, 2nd Ser., vol. VI (1870)
Kebbel, T. E. 'The Agricultural Labourer' (1887)
Kingsley, Charles, 'Alton Locke' (1850), collected edn
——— 'Yeast' (1851) (1889 edn)
Lawes, J. B. and Dr J. H. Gilbert, Report of Experiments on the Growth of Wheat, JRASE, vol. XXIII (1862)
Lawrence, John, 'Farmer's Calendar' (1806)
Lawson, James, 'New Farmer's Calendar' (1827)
Lawson, Wm, and C. D. Hunter 'Ten Years of Gentleman Farming' (1875)
Lewis, Samuel (ed.), 'Topographical Dictionary of England', 2nd edn (1833)
Little, H. J. The Agricultural Labourers, JRASE, 2nd Ser., vol. XIV, pt. II (1878)
Love, Peter, Harvesting in a Bad Season, Prize Essay, JRASE, vol. XXIII (1862)
Lowsley, B. 'A Glossary of Berkshire Words and Phrases' (English Dialect Soc., 1888)
Lubbock, John, 'The Beauties of Nature' (1892)
McConnell, Primrose, 'McConnell's Book of the Farm' (1883)
——— 'The Complete Farmer' (1910), especially Ch. XXXIV, Harvesting Implements
Mackay, Thomas 'History of the English Poor Law', vol. III (Manchester, 1896)
——— 'The English Poor Law' (1899)
——— 'Reminiscences of Albert Pell' (1908)
Main, James, English Agriculture, 'Quart. Journ. Agric.', vol. XIII (1843)
Maine, Lewin George, 'A Berkshire Village' (1866), Stanford-in-the-Vale, Vale of White Horse
Manning, Percy, Some Oxfordshire Seasonal Festivals, 'Folk-Lore', vol. VIII (1897)
Marshall, William, 'A Review and Complete Abstract of the Reports to the Board of Agriculture on the Rural Counties of England' (5 vols., 1817)
Martineau, J. 'The English Country Labourer and the Poor Law in the Reign of Queen Victoria (1901)
Marx, Karl, 'Capital', vol. I., F. Engels (ed.) (1887) (Moscow, 1958 edn), especially Ch. XXXII, Expropriation of the Agricultural Population from the Land
Mason, Violet, Scraps of English folklore - Nineteenth Century Oxfordshire, 'Folk-lore', vol. XL (1929)
Mavor, W. 'General View of the Agriculture of Berkshire' (1813)
Mead, Isaac, 'The Life Story of An Essex Lad' (Chelmsford, 1923)
——— 'A Retrospect. By an Essex Lad' (Chelmsford, 1933)
Milne, William J. 'Reminiscences of an Old Boy 1832 to 1856' (Forfar, 1901)
Mitford, Mary Russell, 'Our Village' (5 vols., 1824-32)
Monk, William J. Witney and its Woollen Manufacture in 'Memorials of Old Oxfordshire', P. H. Ditchfield (ed.) (1903)
Morton, J. C. 'Hand Book of Farm Labour' (1868 edn)
Nicholls, George, On the Conditions of the Agricultural Labourer; with suggestions for its improvement, JRASE, vol. VII (1846)
'Notes and Queries' (1850)
'Parliamentary Gazetteer of England and Wales 1840-1, The' (Glasgow, 1842)
Parker, Angelina, 'Glossary of Words Used in Oxfordshire' (English Dialect Soc., 1876)
——— 'Supplement to Glossary of Words Used in Oxfordshire' (English Dialect Soc., 1881)
——— Oxfordshire Village Folklore 1840-1900, 'Folk-lore', vol. XXIV (1913)
Pearce, W. 'General View of the Agriculture of Berkshire' (1794)
Pell, A. Chert. 'The Making of the Land in England: a Retrospect', JRASE, 2nd Ser., vol. XXIII (1887)

Pitts, J. (printer), 'The Harvester Songster; being a Collection of Choice Songs to be Sung at Harvest Home', printed and sold by J. Pitts (c.1820), Bodleian Library, C.22
Pratt, E. A. 'The Transition in Agriculture' (1906)
Priest, St John, 'General View of the Agriculture of Buckinghamshire' (1813)
Purdy, Frederick, On the Earnings of Agricultural Labourers in England and Wales 1860, JRSS, vol. XXIV (1861)
Pusey, Thomas (MP for Berkshire), 'What Ought Landlords and Farmers to Do? (1851), reprint of article, JRASE, Oxford Union Library, Pamphlet Coll. no. 26
Read, Clare Sewell, On the Farming of Oxfordshire, JRASE, vol. XV (1854)
—— On the Farming of Buckinghamshire, Prize Essay, JRASE, vol. XVI (1855)
Roberts, George, 'The Social History of the People of the Southern Counties of England in Past Centuries (1856)
Rogers, J. E. Thorold, 'Six Centuries of Work and Wages' (1884)
Roscoe, E. S. 'Buckinghamshire Sketches' (1891)
Rose, Walter, 'Fifty Years Ago: by a Native Resident' (192(0))
—— 'Good Neighbours' (1942)
Rugg, KC, A. H. 'The Laws Regulating the Relation of Employer and Workman in England' (1905)
Sampers, William, 'A Digest of... The Results of the Census of England and Wales 1901' (1903)
Sequin, L. G. 'Rural England' (1885)
Simpson, Henry, 'Book of Agriculture' (3 vols., 1875)
Smith, Edward, 'The Peasant's Home 1760' (1875)
Sneyd-Kynnersley, E. M., 'H.M.I.' (1908)
Somerville, Alexander, 'The Autobiography of a Working Man' (1848), Eleanor Eden (ed.) (1862 edn)
—— 'The Whistler at the Plough' (Manchester, 1852)
Spearing, J. B., On the Agriculture of Berkshire, Prize Essay, JRASE, vol. XXI (1860)
Stapleton, Mrs Bryan, 'Three Oxfordshire Parishes, A History of Kidlington, Yarnton and Begbrooke' (Oxford, 1893)
Stephens, James, 'Book of the Farm', James Macdonald (ed.) (1890), 5th edn (1908)
Strutt, J. 'Sports and Pastimes of the People of England' (1801), new edn. W. Hone (1834), J. C. Cox (ed.) (1903)
Stubbs, C. W. 'The Land and The Labourers' (1891)
Sturt, George, 'The Journals of George Sturt', Geoffrey Grigson (ed.) (1941)
—— 'The Journals of George Sturt 1890-1927', vol. I (1890-1904), vol. II (1904-27), edited and introduced E. D. Mackerness (Cambridge, 1967)
Thornton, William Thomas, 'Over-Population' (1846)
Voelcker, Dr Augustus, Experiments with Different Top-Dressing upon Wheat, JRASE, vol. XXIII (1862)
—— The Influence of Chemical Discoveries on the Progress of English Agriculture, JRASE, 2nd Ser., vol. XIV (1878)
Wilson, Henry J., The Dietary of the Labourer, JRASE, vol. XXIV (1863)
Wilson, Rev. John M. 'The Rural Cyclopedia' (3 vols., 1847)
Wodhams, John R. (ed.), 'The Midland Garner' (1884)
Year Book of Facts (1853)
Young, Arthur, 'Annals of Agriculture', vol. X (1788)
—— 'The Farmer's Calendar', 6th edn (1805)
—— 'General View of the Agriculture of Oxfordshire' (1809)
—— 'General View of the Agriculture of Sussex' (1813)

SECONDARY PRINTED SOURCES (INCLUDING ARTICLES UNDER AUTHOR)

Ackland, A. H. Dyke, Roden Buxton and Seebohm Rowntree (Land Enquiry Committee) 'The Land', (1913)
Adams, Leonard P. 'Agricultural Depression and Farm Relief in England 1813-1852' (1932) (1965 edn)

Adamson, J. W. 'English Education 1789-1902' (1930)
Aldcroft, D. H. Communication: The Revolt of the Field, 'Past and Present', no. 27 (1964)
Archer, Fred, 'The Distant Scene', (1967), Gloucestershire
Armitage, W. H. G. 'Four Hundred Years of Education' (Cambridge, 1970)
Ashby, A. W. 'One Hundred Years of Poor Law Administration in a Warwickshire Village', vol. III, Oxford Studies in Social and Legal History, Paul Vinogradoff (ed.) (Oxford, 1912), especially Ch. V, Settlement and Removal
Ashby, M. K. 'The Country School' (1929)
—— 'Joseph Ashby of Tysoe', (1961)
Astor, Viscount and Be. Seebohm Rowntree 'British Agriculture' (1939)
Barraud, E. M. 'Tail Corn' (1948)
Barrett, Hugh (W. H.), 'Early to Rise' (1967)
Barrett, W. H. 'A Fenman's Story' (1965)
Beaumont, Winifred H. 'The Wormingford Story' (Wormingford, 1959)
Bennet, M. K. British Wheat Yields for Four Centuries, 'Economic History', vol. 3, no. 10 (1935)
Blythe, Ronald, 'Akenfield' (1969), Suffolk
Bovill, E. W. 'English Country Life 1780-1830' (1962)
Bracey, H. W. 'English Rural Life' (1959)
Bradley, A. G. 'Round about Wiltshire' (1907)
—— 'Other Days, Recollections of Rural England and Old Virginia 1860-1880' (1913)
—— 'When Squires and Farmers Thrived' (1927)
Brown, A. F. J. 'English History from Essex Sources' (Chelmsford, 1952)
Burnett, John, 'Plenty and Want' (1966), especially pt. II, Ch. 7, Rural England: Romance and Reality, p. 112 ff
—— 'History of the Cost of Living' (1969)
Butlin, R. A. Some Terms Used in Agrarian History, 'Agric. Hist. Rev.', vol. 9 (1961)
Buttress, F. A. 'Agricultural Periodicals of the British Isles 1681 - 1900 and their location' (Cambridge, 1950)
'Century of Agricultural Statistics, Great Britain 1866-1966' (HMSO, 1968)
Chadwick, Owen, 'The Victorian Church' (1964) (1966 edn)
Chambers, J. D. 'The Rural Domestic Industries during the period of transition to the Factory System with Special Reference to the Midland Counties of England' (1962)
—— and G. E. Mingay, 'The Agricultural Revolution 1750-1880' (1966)
Champion, S. G. 'Racial Proverbs' (1938)
Checkland, S. G. 'The Rise of Industrial Society 1815-1885' (1964) (1969 edn)
Clapham, John, 'An Economic History of Modern Britain', vol. 1, The Early Railway Age 1820-1850, 2nd edn rep. with corrections (Cambridge, 1939), vol. 2, Free Trade and Steel 1850-1866 (Cambridge, 1932), vol. 3, Machines and National Rivalries 1887-1914, epilogue 1914-29 (Cambridge 1938) (1951 edn)
Clegg, H. A., Alan Fox and A. F. Thompson, 'A History of British Trade Unions since 1889', vol. 1, 1889-1910 (Oxford, 1964)
Collins, E. J. T. Historical Farm Records, 'Archives', vol. VII (1966)
—— Sickle to Combine, A Review of Harvest Techniques from 1800 to the Present Day (Museum of English Rural Life, Reading, Sept. 1968)
—— Harvest Technology and Labour Supply in Britain 1790-1870, 'Econ. Hist. Rev.', 2nd Ser., vol. XXII, no. 3 (1969)
Coppock, J. T. The Statistical Assessment of British Agriculture, 'Agric. Hist. Rev.', vol. 4, pts. 1 & 2 (1956)
—— 'An Agricultural Atlas of England and Wales' (1964)
Corbett, Elsie, 'A History of Spelsbury' (1962)
Cordeaux, E. H. and D. H. Merry, 'Oxfordshire Bibliography' (Oxford, 1952)
Cornish, J. G. 'Reminiscences of a Country Life (1939)', Berkshire Downs
Court, W. H. B. 'A Concise Economic History of Britain from 1750 to the Present Time' (Cambridge, 1962)
—— 'British Economic History 1870-1914, Commentary and Documents' (Cambridge, 1965)

Cutler, W. H. R. 'A Short History of English Agriculture' (1909)
Darby, H. C. 'The Drainage of the Fens' (Cambridge, 1968)
Darcy, Herbert, 'Poor Law Settlement and Removal' (1910), 3rd edn (1925)
Day, Alice C. 'Glimpses of Sussex Rural Life' (Kingham, n.d., c.1927)
Deane, Phyllis, 'The First Industrial Revolution' (Cambridge, 1965)
—— and W. A. Cole, 'British Economic Growth 1688-1959' (Cambridge, 1962)
Derry, T. K. and T. I. Williams, 'A Short History of British Technology' (Oxford, 1960)
Docking, J. W. Victorian Schools and Scholars, Coventry and North Warwickshire History Pamphlets, no. 3 (1967)
Drescher, L. 'The Development of Agricultural Production in Great Britain and Ireland', Manchester School, vol. XXIII (1955)
Driver, Cecil, 'Life of Richard Oastler, Tory Radical' (Oxford, 1946), especially Ch. XXII, The Problem of the Poor Law
Dunbabin, J. P. D., The Revolt of the Field: The Agricultural Labourers' Movement in the 1870's, 'Past and Present', no. 26 (Nov. 1963)
—— The Incidence and Organisation of Agricultural Trades Unionism in the 1870's, 'Agric. Hist. Rev.', vol. 6 (1968)
Dunlop, J. 'The Farm Labourer' (1913)
Eggar, J. Alfred, 'Remembrances of Life and Custom in Gilbert White's, Cobbetts' and Charles Kingsley's Country' (1924)
Eland, G. 'In Bucks: Old Works and Post Days in Rural Buckinghamshire' (Aylesbury, 1923)
Ernle, Lord, 'English Farming Past and Present', 6th edn (1961)
Evans, E. E. 'Irish Heritage' (Dundalk, 1942) (7th Impression 1958)
—— 'Irish Folkways' (Dundalk, 1957), especially Ch. XII, Harvest
Evans, G. E. 'Ask the Fellows who cut the Hay' (1956)
—— 'The Farm and the Village' (1969)
—— 'Where Beards Wag All. The Relevance of Oral Tradition' (1971)
Fairlie, Susan, The Nineteenth-Century Corn Law Reconsidered, 'Econ. Hist. Rev.', 2nd Ser., vol. XVIII, no. 3 (Dec. 1965)
—— The Corn Laws and British Wheat Production 1829-76, 'Econ. Hist. Rev.', 2nd Ser., vol. XXII, no. 1 (Apr. 1969)
Finch, William Coles, 'Life in Rural England' (1928), Kent
Fitzgerald, Kevin, 'Ahead of their Time: A Short History of the Farmers' Club, 1842-1967' (1968)
Fletcher, T. W. The Great Depression in English Agriculture 1873-1896, 'Econ. Hist. Rev.', 2nd Ser., vol. XIV (1961)
Frazer, J. G. 'Spirits of the Corn and of the Wild' vol. III, The Golden Bough (1912)
Fussell, G. E. English Agriculture: From Arthur Young to William Cobbett, 'Econ. Hist. Rev.', vol. VI, no. 2 (1936)
—— Home Counties Farming 1840-1880, 'Econ. Journ.', vol. LXVII, no. 227 (1947)
—— 'From Tolpuddle to T.U.C.: A Century of Farm Labourers' Politics' (Slough, 1948)
—— 'The English Rural Labourer: Home Furniture, Clothing from Tudor to Victorian Times' (1949), especially Ch. IX
—— 'The Farmer's Tools A.D. 1500-1900' (1952)
—— and F. R. Fussell, 'English Country Woman A.D. 1500-1900' (1953)
—— and F. R. Fussell, 'The English Countryman 1500-1900' (1955)
—— 'Farming Technique from Prehistoric to Modern Times' (Oxford, 1966)
—— 'A Bibliography of His Writings on Agricultural History' (Reading Univ., 1967)
George, Dorothy, 'England in Transition, Life and Work in the Eighteenth Century' (1931) (rep. 1962)
Geraint Jenkins, J. 'The English Farm Waggon' (Reading, 1961)
Gonner, E. K. 'Common Land and Inclosure' (1912) (rep. 1966, with new intro. by G. E. Mingay)
Green, F. E. 'The Tyranny of the Countryside' (1913) (1921 edn)
—— 'A History of the English Agricultural Labourer 1870-1920' (1920)

Green, M. C. F. 'The Village Labourer' (1927)
Grigson, Geoffrey, 'The English Year' (1967)
Groves, R. 'Sharpen the Sickle! the History or the Farm Workers' Union' (1949)
Habakkuk, H. J. The Economic Growth of Modern Britain, 'Journ. of Econ. Hist.' vol. XVIII (1958)
Hall, A. D. (Sir Daniel), 'A Pilgrimage of British Farming' (1913), a description county by county in 1910
Hammond, J. L. and Barbara Hammond, 'Lord Shaftesbury' (1923) (1939 edn), especially Ch. XIII
Hammond, N. K. 'Golden Berkshire' (1967)
Handley, James E. 'The Irish in Scotland 1798-1845' (Cork, 1943)
—— 'The Irish in Modern Scotland' (Cork, 1947)
—— 'The Agricultural Revolution in Scotland' (Glasgow, 1963)
Hargreaves, Barbara (ed.), 'The Diary of a Farmer's Wife' (1964)
Harman, H. 'Buckinghamshire Dialect Glossary' (1929)
Hart, A. Tindall, 'The Country Priest in English History' (1959)
Harvey, Barbara F. (Ed.), 'Customal (1391) and Bye-Laws (1386-1540) of the Manor of Islip' (Oxfordshire Record Soc., 1959)
Hasbach, W. 'The English Agricultural Labourer' (1908)
Healy, M. J. R. and E. L. Jones, Wheat Yields in England 1815-57, JRSS, Ser. A, vol. 125 (1962)
Hobsbawm, E. J. The British Standard of Living 1790-1850, 'Econ. Hist. Rev.', 2nd Ser., vol. X, no. 1 (1957)
—— and R. M. Hartwell, The Standard of Living During the Industrial Revolution: A discussion, 'Econ. Hist. Rev.', 2nd Ser., vol. XVI (Aug. 1963)
—— 'The Age of Revolution' (1964)
—— 'Industry and Empire' (1968)
—— and George Rude, 'Captain Swing' (1968)
Holdenby, Christopher, 'Folk of the Furrow' (1911)
Hoskins, W. G. 'The Midland Peasant' (1957)
Humphreys, A. L. 'Berkshire Book of Song' (1935)
Hunt, E. H. Labour Productivity in English Agriculture 1850-1914, 'Econ. Hist. Rev.', vol. XX (1967)
Jackson, John Archer, 'The Irish in Britain' (1963)
Jebb, L. 'The Labourer' (1907)
Jobson, A. 'Suffolk Yesterdays' (Heath Cranton, 1944)
John, A. H. The Course of Agricultural Change, 'Studies in the Industrial Revolution', L. S. Pressnell (ed.) (1960)
Johnson, J. H. Harvest Migration from Ireland, 'Institute of British Geography Transcript', 41 (1967)
Johnson, Marion, 'Derbyshire Village Schools in the Nineteenth Century' (Newton Abbot, 1970)
Jones, E. L. English Farming Before and During the 19th Century, 'Econ. Hist. Rev.', vol. XV (1962)
—— The Agricultural Labour Market in England 1793-1872, 'Econ. Hist. Rev.', 2nd Ser., vol. XVII (1964)
—— 'Seasons and Prices. The Role of Weather in English Agricultural History' (1964)
—— (ed.) 'Agriculture and Economic Growth in England 1650-1875' (1967)
—— (ed.) 'Land, Labour and Population in the Industrial Revolution. Essays presented to J. D. Chambers' (1967)
—— Agricultural Origins of Industry, 'Past and Present' (1968)
—— 'The Development of English Agriculture 1815-1873' (1968)
—— and E. J. T. Collins, The Collection and Analysis of Farm Record Books, 'Journ. of the Soc. of Archivists', vol. III (1965)
Julyan, H. E. 'Rottingdean and East Sussex Downs and Villages' (Lewes, 1940)
Kerr, Barbara M. Irish Seasonal Migration, 'Irish Historical Studies', vol. III (1942-3)
—— 'Bound to the Soil; a Social History of Dorset, 1750-1918' (1968)
Kitchen, Fred, 'Brother to the Ox' (1940), Yorkshire
Lee, Norman E. 'Harvest and Harvesting' (Cambridge, 1960)

McCann, W. P. Elementary Education in England and Wales on the Eve of the 1870 Education Act, 'Journal of Educational Administration and History, Univ. of Leeds', vol. II, no. 1 (Dec. 1969)
McClatchey, D. 'Oxfordshire Clergy 1777-1869, A Study of the Established Church and of the Role of its Clergy in Local Society' (1960)
Martin, E. W. 'The Countryman's Chap Book' (1949)
—— 'The Secret People. English Village Life after 1750' (1954) (1956 edn)
Merrman, John R. H. 'A History of the Church of England' (1953)
Minchinton, W. E. (ed.), 'Essays in Agrarian History' (2 vols., 1967)
Moore, D. C. The Corn Laws and High Farming, 'Econ. Hist. Rev.', 2nd Ser., vol. XVIII, no. 3 (Dec. 1965)
Moreau, G. E. 'The Departed Village' (1968), Britwell Salome, Oxon
Morgan, David H. Oxfordshire Travelling Harvesters, 'The Oxford Magazine', 22 May 1970
Morgan, E. Victor, 'The Study of Prices and the Value of Money' (Historical Association, 1950)
Morley, Edwin, 'A Descriptive Account of East Hendred' (Oxford, 1969)
Morris, M. C. F. 'The British Wormkan' (1928)
Nettel, R. 'A Social History of Traditional Song' (1969)
Olson Jnr, Mancur and Curtis C. Harris Jnr, Free Trade in 'Corn', A Statistical Study of the Price of Wheat in Great Britain, 'Quart. Journ. of Econ.', vol. LXXIII (1959)
Ojala, Eric M. 'Agricultural and Economic Progress', Univ. of Oxford Agric. Econ. Research Institute (Oxford, 1952)
Orr, John, 'Agriculture in Oxfordshire', (Oxford, 1916)
—— 'Agriculture in Berkshire' (Oxford, 1918)
Orwin, Christobel S. and B. Felton, A Century of Wages and Earnings in Agriculture, JRASE, 4th Ser., vol. XVII (1931)
—— and E. H. Whetham, 'History of British Agriculture 1846-1914' (1964)
Parry, Major Gambier-, 'The Spirit of the Old Folk' (1913)
Payne, F. G. The Retention of Simple Agricultural Techniques, 'Gwerin II' (1959)
Pedder, D. C. 'Where Men Decay; A Survey of Present Rural Conditions' (1908)
Percival, John, 'Wheat in Great Britain' (1948)
Perkins, Harold, 'The Origins of Modern English Society 1780-1880' (1969)
Phelps, Gilbert, 'A Survey of English Literature' (1965)
Pike, Royston E. 'Human Documents of the Victorian Golden Age' (1967)
Plaisted, A. H. 'The Manor and Parish Records of Medmenham, Bucks' (1925)
Radzinowicz, Leon, 'A History of English Communal Law', vol. 4 (1968)
Randell, Arthur, 'Fenland Memories' (1965)
—— 'Sixty Years a Fenman', E. Porter (ed.) (1966)
Raven, C. E. 'Science, Religion and the Future' (Cambridge, 1968)
Redford, Arthur, 'Economic History of England 1760-1860' (1960)
—— 'Labour Migration in England 1800-1850' (Manchester, 1926), 2nd edn W. Chaloner (Manchester, 1964)
Rew, R. H. 'An Agricultural Faggot' (1913), studies in the agricultural production
Rostow, W. W. Business Cycles Harvest and Politics 1790-1850, 'Journ. Econ. Hist.', vol. I (1941)
—— 'British Economy of the Nineteenth Century' (1948)
Rowntree, B. Seebohm and M. A.-I. Wendell, 'How the Labourer Lives' (1913)
Ryder, M. L. The Size of Haystacks, 'Folk Life', vol. 7 (1969)
Sage, J. 'The Memoirs of Josiah Sage' (1951)
Saul, S. B. 'The Myths of the Great Depression in 1873-1896' (1969)
Saville, John, 'Rural Depopulation in England and Wales 1851-1951' (1957)
—— Primitive Accumulations and Early Industrialization in Britain, 'Socialist Register' (1969)
Seebohm, M. E. 'The Evolution of the English Farm' (1952)
Selley, Ernest, 'Village Trade Unions in Two Centuries' (1919)
Sharpe, Frederick, 'The Church Bells of the Deanery of Bicester' (Brackley, 1932)

—— 'The Church Bells of Oxfordshire' (4 vols., Oxfordshire Record Soc., Oxford, 1945-52)
Sibbit, Christine, 'Balls, blankets, baskets and boats, a survey of crafts and industries in Oxfordshire', Oxford City and County Museum, no. 1 (1968)
Simon, Daphne, Master and Servant in 'Democracy and The Labour Movement', John Saville (ed.) (1954)
Storr-Best, Lloyd, 'Varro on Farming' (1912)
Sturgess, R. W. The Agricultural Revolution on the English Clays, 'Agric. Hist. Rev.', (1966)
Sturt, Mary, 'The Education of the People' (1969), especially Ch. XIV
Styles, Philip, The Evolution of the Law of Settlement, 'Univ. of Birmingham Historical Journal', vol. IX (1963)
Syddow, C. W. The Mannhardtian Theories about the Last Sheaf and the Fertility Demons, 'Folk-lore', vol. XLV (1934)
Thirsk, Joan, 'English Peasant Farming. The Agrarian History of Lincolnshire from Tudor to Recent Times' (1957)
—— and Jean Imray (eds.) 'Suffolk Farming in the Nineteenth Century', Suffolk Record Society, vol. I (Ipswich, 1958)
Thompson, E. P. 'The Making of the English Working Class (1968)
Thompson, Flora, 'Larkrise to Candleford' (1965 edn)
Thompson, F. M. L. 'English Landed Society in the Nineteenth Century' (1963)
—— The Second Agricultural Revolution 1815-1880, 'Econ. Hist. Rev.', vol. XXI, no. 1 (Apr. 1968)
Timmer, C. Peter, The Turnip, The New Husbandry and the English Agricultural Revolution, 'Quart. Journ. Econ.', vol. LXXXIII, no. 3 (Aug. 1969)
Trinder, Barrie S. Banbury's Poor in 1850, Banbury Historical Society (1966), reprinted from 'Cake and Cockhorse', vol. 3, no. 6 (1966)
Smith, R. Trow, 'English Husbandry' (1951)
—— 'Society and the Land' (1953)
'Victoria County History of Berkshire', vol. I (1906), vol. II (1907)
'Victoria County History of Buckinghamshire', vol. I (1905), vol. II (1908)
'Victoria County History of Oxfordshire', vol. II (1907)
Vulliamy, A. F. 'The Law of Settlement and Removal' (1906)
Warren, C. Henry, 'The English Countryman' (1955), semi-autobiography of Mark Rushton
Watson, James A. Scott and May Elliot Hobbs, 'Great Farmers' (1951)
Webb, S. and B. Webb, 'English Poor Law History', pts. 1 & 2, vol. VII, English Local Government (1963 edn)
—— (eds.) 'The Break-up of the Poor Law: being Pt. I of the Minority Report of the Poor Law Commission. The Public Organisation of the Labour Market, being Pt. II...' (1909)
Whistler, Lawrence, 'The English Festivals' (1947)
Williams, Alfred, 'A Wiltshire Village' (1912)
—— 'Folk Songs of the Upper Thames' (1922)
—— 'Round About the Upper Thames' (1922)
Williams, M. I. Seasonal Migration, 'Credigion', vol. III (1957)
Woods, K. S. 'The Rural Industries Round Oxford' (1921)
Wright, A. R. and T. E. Lones, 'British Calendar Customs, Vol. I. Movable Feasts', Folk-lore Society, vol. XCVII (1936)
—— 'British Calendar Customs Vol. III England', vol. CVI (Folk-lore Society, 1940)
Wright, Philip A. 'Old Farm Implements' (1961)

THESES

Collins, E. J. T. Harvest Technology and Labour Supply, PhD Thesis, Nottingham Univ., May 1970
Horn, Pamela L. R. Agricultural Labourers' Trade Unionism in Four Midland Counties 1860-1900; Leicestershire, Northamptonshire, Oxfordshire and Warwickshire, PhD Thesis, Univ. of Leicester, 1968

Jones, Gareth Stedman, Some Social Consequences of the Casual Labourer Problem in London 1860-1890 with particular reference to the East End, PhD Thesis Oxford Univ., 1969

INDEX